T0213748

Virtualization Techniques for Mobile Systems

David Jaramillo
IBM
Lake Worth, FL, USA

Ankur Agarwal
Florida Atlantic University
Coral Springs, FL, USA

Borko Furht
Department of Computer Science
& Engineering
Florida Atlantic University
Boca Raton, FL, USA

ISBN 978-3-319-38180-0   ISBN 978-3-319-05741-5 (eBook)
DOI 10.1007/978-3-319-05741-5
Springer Cham Heidelberg New York Dordrecht London

Printed on acid-free paper

Springer is part of Springer Science+Business Media (www.springer.com)

Contents

List of Figures

List of Tables

URI	Universal resource identifier
URL	Universal resource locator
VM	Virtual machine
VNC	Virtual network computing
VPN	Virtual private network

Chapter 1
Introduction

In a short span of time, mobile phones have matured from limited oversized gadgets to small powerful feature packed smart phones. Early feature phone devices offered little more function than basic email, calendar and contacts. Due to advances in semiconductor technology, these devices started offering the computation power comparable to a slightly older generation computer and made them capable of running a wide variety of user installed applications. With the emergence of apps, there came the increased importance of protecting the data on the device, much of which was of a personal nature. As adoption in the general public grew, people wanted to use their smartphones in the workplace; Enterprises began to embrace the new platforms under the banner of Bring Your Own Device (BYOD) and set about discovering ways to use them in enhancing the overall productivity and communication among team members. This brought about the challenge of how to manage devices not owned by the company yet still provide methods to securely access enterprise information in a way that complied with company policy, which in turn led to enterprise level device management and policy enforcement.

1.1 Motivation

Enterprises are being faced with supporting employee owned devices since they no longer wish to incur the cost of paying for devices and phone/data services. Employees are buying their own devices that have enterprise capability and want to connect to the enterprise network so that they can do their work with greater flexibility and freedom. However, the employees also don't want to give up user experience and freedom at the cost of complex IT security policies. There are a number of components such as device choice, security models, liability, user experience, privacy and economics that need to be considered in delivering a BYOD offering that works well for the user and the enterprise. In order to achieve adoption and sustainability from the BYOD community, in the enterprise from the BYOD

D. Jaramillo et al., *Virtualization Techniques for Mobile Systems*,
Multimedia Systems and Applications, DOI 10.1007/978-3-319-05741-5_1,
© Springer International Publishing Switzerland 2014

community, mobile virtualization is rapidly becoming a very attractive choice because it offers flexibility and addresses the concerns over privacy of personal data while also delivering the security requirements of the Enterprise.

The work in this book addresses the separation between personal and enterprise, presents and discusses the development and research that has been done in the last few years to address this mobile virtualization and presents a comparison analysis of these various solutions. Investigation is done with technologies such as Hypervisors which are evolutions from the desktop and server environments that have special CPU and memory requirements, as well as other technologies like application containers which leverage operating system and application permissions to provide separation between the personal and enterprise. Lastly, more traditional separation has been through the use of Device Management Policies to dictate the persona of applications and data.

1.2   Enhancing User Productivity

In order to achieve adoption and sustainability in the enterprise, mobile virtualization is rapidly becoming a very attractive choice because it addresses the flexibility while preserving the privacy concerns from the user and it also delivers the security requirements for the enterprise. On the other side of the ecosystem, the device makers and carriers will benefit from mobile virtualization because they are able to more easily replicate the features found in various devices and also deliver more features at a lower cost. Allowing BYOD devices in the enterprise is more than just giving in to employee preferences, but it means to put policies in place that govern how devices will be used and how they will be managed while maintaining end user flexibility.

1.3   Improving System Security

Mobile device security in a BYOD environment is critical because you want to protect the user's device from being compromised from internet attacks which could lead to loss of personal and enterprise data. Various forms of protection are built in as part of the Operating System like device encryption, password policies, remote lock, remote selective/full wipe and more, whereas additional security is supplemented via third party applications that take advantage of Device Administration API's that are built into the Operating System to provide protection against malware, application scanning upon installation and more.

Device security is key in maintaining data integrity, but a practice amongst users that is becoming a more widely seen is that of device rooting which can be a very serious security exposure for enterprises [Grum01]. Essentially, rooting a device is the process of granting privileged access in the Operating System to manage the device, thereby allowing the user to overcome the limitations that are implemented to safe guard the device hardware. Rooting installs a new Superuser that can affect application permissions by giving them more access to allow them to have system level privileges. Common examples of this are: to enable Hotspot WiFi capabilities thereby bypassing carrier restrictions, enabling the device to do VNC screen sharing or enabling other application stores like Cydia and others that allow for unrestricted applications to be installed.

Operating Systems like Android implement various mechanisms to maintain system security and reduce the vulnerabilities in the system. One mechanism is Application Sandboxing which enforces each application to run in its own process and assign a specific user and group-id to manage access to files by process ownership. Complementing this mechanism is Application Permissions, where a declarative permission model is used to enforce restrictions on specific operations that application wishes to perform or device resources the application wishes to access. Through the combination of these mechanisms, the system is made fairly secure; however there are limitations that make the system susceptible to data leakage over the internet as with with social networking applications [KhNa01].

1.4 Contributions

The work presented focuses on the study of various techniques in mobile virtualization and how it is applied in the mobile enterprise. These contributions can be summarized as follows:

1. A comparative study of various mobile virtualization techniques and technologies has been performed.
2. A case study is presented of a mobile virtualization pilot that was conducted where one of the technologies tested and surveyed with a group of mobile users within an enterprise. The pilot survey results were then analyzed and presented.
3. An innovative reference architecture for mobile virtualization has been developed and presented, which is suitable for mobile hybrid based applications.
4. The reference architecture has been tested and compared with other mobile virtualization techniques using a performance benchmark. The results were then analyzed through the use of statistical analysis to help identify the performance impacts in each of the virtualization technologies across the various platforms that were tested.

With the work presented here, deployment plans and strategies can be developed more intelligently for integrating the growing BYOD community into the enterprise mobile ecosystem. The number of BYOD users is growing rapidly due to the rapid adoption of smartphones by consumers and with the research presented here, enterprises are able to make much more intelligent decisions in the choices of technologies to adopt and deploy.

1.5 Proposed Methodology/Innovation

The proposed work presented here is an innovate approach of a Mobile Virtualization Computing Model for mobile devices. For designing such a system, we have defined a concept of a layered Mobile Computing Security model that will integrate various security components into the appropriate layered architecture. Layered architecture is an effective way of dividing and conquering a problem. It provides a way to separate the concerns of various domains, thus, providing an effective solution to a domain specific problem. It can be also argued that this approach will provide a flexible and scalable solution in addressing inter-domain specific issues.

We propose a layered system design that will be based on subsystem components that can be added or removed to provide dynamic adaptability for the user profiles depending on the user's domain environment. This system will adapt itself in cases where any one component becomes enabled in order to maintain system integrity, enable/disable components depending on the location of the user, adjust performance utilization based on the task being performed and provide authentication methods will adapt based on the user domain profile.

A standard development process as described in Fig. 1.1 will be used to develop the system being proposed. Initially, the requirements and specification will be collected. The system solution will then be modeled and analyzed through the use of use case scenarios. This will then be followed by a system design process that will be verified through several use case scenarios, at which point a prototype will be developed where the model will be tested.

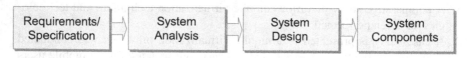

Fig. 1.1 Mobile software development process

Chapter 2
Mobile Virtualization Technologies

In order to achieve increased adoption and sustainability "bring your own device" (BYOD) schemes within the enterprise, mobile virtualization is rapidly becoming a very attractive choice because it provides both employee and enterprise with flexibility while addressing the privacy concerns of the user and meeting the organizations security requirements. Allowing BYOD devices in the enterprise requires policies in place that govern how devices will be used and how they will be managed while maintaining end user flexibility. A number of technologies for mobile virtualization have been developed over the last few years which range from sophisticated mobile device policy management, to hypervisors and container based separation [JaKa01].

2.1 Mobile Virtualization via Device Management Policies

Mobile Separation can be achieved through the use of IT Security Policies that are managed by a Mobile Device Management (MDM) system. This type of approach is done by a server based management approach that lets the IT managers enforce policies across the user base which can be applied to the whole community or on a group basis so that it can be customized to the level of security required per group. For example, executives can have a stricter set of policies to include device encryption and a general group that allows for limited email, calendar and contacts support. There are many MDM's available now in the market and most of them support the major Operating Systems like iOS, Android, RIM and Windows. Some examples of these that provide a wide range of security policies that manage separation of applications and data are BlackBerry Enterprise Server, MobileIron and Good Technology amongst several more [Grum02].

D. Jaramillo et al., *Virtualization Techniques for Mobile Systems*,
Multimedia Systems and Applications, DOI 10.1007/978-3-319-05741-5_2,
© Springer International Publishing Switzerland 2014

Installation of the hypervisor is done on top of the guest OS because it is just like any other application. Performance of the guest OS is heavily dependent on the host OS. Furthermore, any compromise of the host OS will render the guest OS inoperative as well [CrSo01].

Hypervisor technologies have a downside in that they are required to work directly with OEMs which takes longer and there are fewer smartphones today that can support hardware level virtualization [Bran01], however, this will change over time as the ARM Cortex-A7, A15 and similar processors are incorporated into more mobile smartphones/tablets and also as standards like Virtualization Management Object (VirMO) proposed by Red Bend in the Open Mobile Alliance Device Management (OMA DM) Working Group [Red01].

2.2.1 KVM on ARM

Kernel based Virtual Machine (KVM) is a virtualization infrastructure for the Linux kernel that supports native virtualizations on processors with hardware virtualization extensions. KVM/ARM is a virtualization solution for ARM processor based devices that can run virtual machines with nearly unmodified operating systems. Since the ARM CPU processor is not virtualizable, KVM/ARM uses a lightweight paravirtualization [DaNi01] via a script-based method to automatically modify the source code of an operating system kernel to allow it to run in a virtual machine. This lightweight paravirtualization is architecture specific but operating system independent as seen on the above figure [DaNi01]. These changes in the guest OS kernel are made so that it can take care of sensitive non-privileged instructions (Fig. 2.2) by doing the trap and emulate methods which are then handled by an interrupt handler which then emulates the appropriate functionality [Rama01].

2.2.2 Xen Hypervisor on ARM

Xen is an open-source hypervisor that allows for multiple operating systems to safely share the hardware via resource management without sacrificing performance or functionality [BaDr01]. Figure 2.3 illustrates a basic Xen configuration where the hypervisor consists of a small layer on top of the physical hardware. It implements virtual resources such as vMemory, vCPU, event channels and shared memory, and it controls the assignment of I/O devices to VMs. The user domains—DomUs are started by the Dom0 and they can run any paravirtualized operating systems like Linux and others. These guest OSs have minimal changes where privileged operations are changed to calls to the hypervisor [SaVa01].

Xen has been ported to the ARM architecture [Xen01] used for secure mobile phones supporting enhanced security features for mobile devices with mandatory

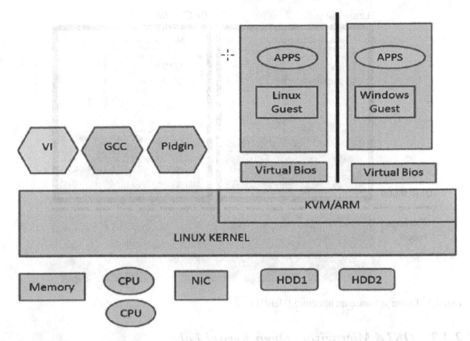

Fig. 2.2 KVM overview [Rama01]

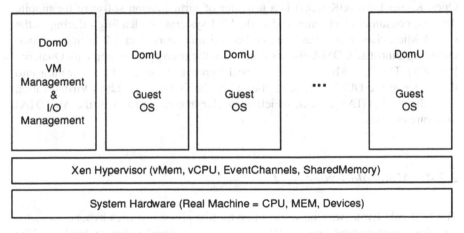

Fig. 2.3 Xen hypervisor [Xen01]

access control through access control models and a secure boot process which is designed to detect any alterations of the VM during the bootstrap process. Samsung has taken a keen interest in this project and has developed a version that supports ARMv5, ARMv6 and ARMv7 processors, the later one supporting the new virtualization extensions [Morg01].

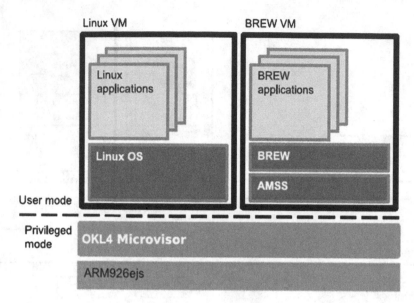

Fig. 2.4 Evoke software architecture [Heis01]

2.2.3 OKL4 Microvisor: Open Kernel Labs

Open Kernel Labs (OK Labs) is a provider of virtualization software for mobile devices, consumer electronics and embedded systems. Its leading offering is the OKL4 Microvisor which has been embedded into more than 1.2 billion devices, including almost all CDMA phones because of the strong partnership with Qualcom [CrSo01]. The Okl4 Microvisor is a type 1 hypervisor that can be either built into the device at the OEM level or applied after the fact via OK Lab's Virtualization Over the Air (VOTA) process, which is similar in concept to Over the Air (OTA) firmware updates.

2.2.4 Motorola Evoke AQ4

The Motorola Evoke was the world's first mobile phone that uses hypervisor virtualization, implemented using OKL4 as the core virtualization technology. The requirements for the phone were such that they only could be met by a design based on virtualization. This was accomplished by using a phone with a specific price point based on a single core design using an ARM9 core, a user interface running on Linux (OS), the baseband stack running outside of Linux, and components from BREW UI framework were re-used (Fig. 2.4).

Evoke's software architecture is shown in the diagram below, where there are two virtual machines running on top of the OKL4 Microvisor which interact via the OKL4 message-passing IPC as well as through shared memory. The complete

Linux system is running de-privileged and the AMSS/BREW baseband stack/OS are both running in user mode. Due to the high performance of the OKL4 Microvisor, the virtualization overhead was kept to almost unnoticeable levels which in many respects were better than what was achievable with running native Linux.

As a result, Motorola produced an attractive device with a snappy user interface that was even more responsive than other non-virtualized phones, even those which were based on more powerful ARM11 processors [Heis01].

2.2.5 VMWare

VMware is committed to bringing virtualization to the mobile handset and two manufacturers—LG and Samsung have announced their support for this solution on Google Android [Thom01]. At VMworld 2011, VMWare announced VMware Horizon Mobile Manager, previously known as VMware Mobile Virtualization Platform (MVP) which allows Android phones to use virtual machine technology to run a second instance of Android, very much the same way virtualization works on servers and desktops. This solution basically has two separate phones running on one device, and can switch from the personal one to the corporate one by clicking a "work phone" icon. By isolating the employee's work environment from their personal environment and providing IT managers a Web-based management console to control what employees can do on the work portion of the phone, the user can have a more relaxed personal experience while the enterprise can have a manageable and secure environment [VMTN1] (Fig. 2.5).

Having two environments on the same device doesn't really make things more complex for the user because common functions such as receiving a call are active regardless of which environment is active. VMware says performance impact will be minimal and that the offering will work on both single and dual-core processors. Google Android was picked for the development of this platform largely due to the flexibility of it being open source. Initially, LG signed on to the project last December and now recently, Samsung has joined in providing a wider variety of devices such as the Galaxy S II phones and Galaxy tablets. In the future, more devices and other manufacturers are to be supported according to press announcements [Zieg01].

It is possible today to do many work activities on Android phones and tablets, but the real issue is how to ensure that the enterprise doesn't get affected by malware, viruses or losing data when the device is lost. This is why VMware's Horizon Mobile is very attractive because the enterprise phone is not affected by any malicious software and can be managed by IT administrators. If the device is lost, it can be remotely locked or wiped. Furthermore, the virtual phone can be provisioned with standardized or custom templates and also push out application updates over the air. User policies can be defined that can restrict what functions/features can be used and configure security features such as device lockup timeouts and passwords. From the employee's perspective, they are happy because they can now use their favorite personal device to do their work as well as use their favorite personal applications without having to carry two different devices [Whit01].

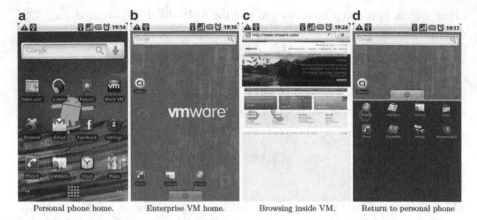

a	b	c	d
Personal phone home.	Enterprise VM home.	Browsing inside VM.	Return to personal phone

Fig. 2.5 MVP personal/enterprise screen shots [VMTN1]

for a physical in-person installation. Furthermore, the IT administrator can make adjustments, add/remove applications, push down new templates without having to reload the device. (4) Review health of the deployment and vital stats from a dashboard. (5) Application management with the ability to push applications over-the-air to the work phone from the application catalog. Allows the add/remove of apps and dynamically pushing the updates to the device. Multiple application versions are allowed. (6) Lock or wipe to de-provision the work phone which allows the ability to lock or wipe a device when the device is lost or sometimes the user might want to just reload the enterprise side [VMTN1].

On the device side, VMware Horizon Mobile Platform is built around a type 2 mobile hypervisor that is based on a lightweight paravirtualization technique for ARMv7 cores that is aimed at minimizing the total system complexity. A series of device and platform virtualization approaches for storage, networking and telephony are implemented which are key in enabling the performance, reliability and security of the system. Finally, this hypervisor is applied to the virtualization of the Android operating system, allowing it to run both the guest and host environments [BaBu01].

VHMM differentiates from previous approaches of system virtualization on the ARMv4-7 architectures that have entailed some form of core paravirtualization like Xen on Arm [SaVa01] by employing a distinct shallow paravirtualization approach that requires only the identification and replacement of sensitive instructions [BaBu01].

2.2.6 Red Bend: vLogix Mobile

Red Bend acquired VirtualLogix, a provider of mobile device virtualization solutions which it delivered to semiconductor vendors, OEMs, ODMs' service providers, and systems integrators. Over one million devices shipped with VirtualLogix's type

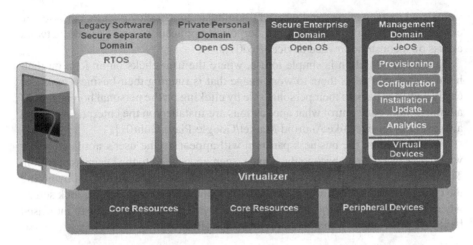

Fig. 2.6 vLogix software architecture [CrSo01]

1 hypervisor technology embedded on devices like the Acer beTouch E110, E120, and ch E130 models; HTC Tianyi; KTouch W606; and CoolPad Yulong W711. Red Bend adopted this technology to fit into it's broader portfolio and also saw it could use its strength, which is its ability to partition secure software domains that can be managed separately [CrSo01].

This offering supports processors based on the ARM Cotex-A15 and Cortex-A7 cores in single and multi-core configurations. Red Bend is enabling device manufacturers to take advantage of this latest family of cores without the need to modify an existing high-level operating system (HLOS). vLogix Mobile performs management of the chipset's multi-core architecture in the Virtualizer, allowing the HLOS to remain as is. Users of this technology benefit by not having to redesign, redevelop and revalidate existing software to support new OS configurations [Red01] (Fig. 2.6).

The solution is comprised of the Virtualizer which is a type-1 hypervisor that runs together with suite of software modules that are configurable according to the desired deployment. The Virtualizer runs on the host hardware, also known as a Bare Metal Hyper, and schedules access to the shared hardware services like file systems, serial lines and network interfaces amongst the virtualized operating environments. System resources like RAM and persistent storage like flash memory are partitioned and allocated according to the performance demands of each of the operating domains or virtual machines. Hardware resources like CPU, clock and memory management unit are also virtualized for each of the guest operating systems and the access to the actual hardware is allocated by the Virtualizer [Red01].

Virtual devices are provided for each of the guest operating systems that access system resources like screen, 2D/3D hardware acceleration, multimedia acceleration, Wi-Fi, GPS and other input/output devices. This provides a secure separation in each of the OS/domains to the features of each device without the physical access to the device itself. The Domains are virtual machines that are designed to run virtualized images of Android where IT administrators deploy over the air a secure enterprise environment that can run corporate applications and provide network access on a consumer owned device [Red01].

Red Bend's solution is simple to use, where the user clicks on an icon on their home screen that takes them to work image that is running their business apps and easily switches back to their personal side by clicking on the personal home icon. IT administrators can control what applications are installed on the enterprise side and also exclude services like Android Market/Google Play [Gohr01].

Notifications in the business partition will appear on the user's notification bar where they can then go back to the home screen and switch to the business partition. The difference between Type 1 and Type 2 hypervisors can be visualized by noting the change from one platform to another when one is done at the phone lock screen where the other one is done at the application level, thereby providing an increased hardware level security and better performance [Bran01].

2.2.7 Cells

Cells' virtualization architecture is used for enabling multiples smartphones that run simultaneously on the same physical cell phone in an isolated manner. This architecture follows a usage model where there is one foreground virtual phone and multiple background virtual phones. A device namespace mechanism and device proxies are integrated with a lightweight operating system virtualization to multiplex phone hardware across multiple virtual phones while providing native hardware device performance. The platform includes a fully accelerated 3D graphics, complete power management features, and full telephony functionality with separately assignable telephone numbers and caller ID support. A prototype was implemented of Cells that supports multiple Android virtual phones on the same phone. Performance results demonstrated that Cells imposed only modest runtime and memory overhead, worked seamlessly across multiple hardware android smartphone devices and transparently runs Android applications at native speed without any modifications [AnDa01].

2.2.8 Cellrox

Cellrox is an Android based technology that has roots on the Cells project [Cell01] and it is not considered to be a complete OS virtualization because the virtualization is limited to the user space, creating multiple personas that share a common Linux kernel. Since the kernel is shared, all of the personas have to be a similar version of Android. Cellrox provides the same look and feel as the original operating system in order to provide a common user experience between the personal and work personas. Multi-tasking between the various personas works much the same way as how multi-tasking works between different Android apps where only one persona is in the foreground at any given time and the apps running in other personas run just like any other background app. Cellrox has the ability to support more than two personas which gives the ability to have a multi-tear level of locking down applications. For example, the personal persona can be very relaxed, while a work persona has semi-restricted environment with standard security policies for the enterprise and third persona could be a completely locked down environment for highly confidential applications. One of the outstanding features that Cellrox features is the ability to have shortcuts of the applications from the different personas such that the user doesn't have to jump between the different personas to find the application which is the case in most of the separation technologies [Madd02].

Usability is a key component in Cellrox. Only one persona is in the foreground while the others run in the background. The user can easily switch between the personas by using a custom key-combination to cycle through the personas or by swiping up and down on the home screen of a persona and each persona has an application icon that can be launched to see a complete list of available personas to select from. For security purposes, the system can be configured such that a no-auto-switch option will prevent background personas from being switched to the foreground without explicit user consent, preventing a background persona from appearing unexpectedly. An auto-lock feature can be enabled that will require the user to unlock the persona using a pass code or gesture whenever a persona switches from the background to the foreground. CellRox's technology was evaluated through an experimental study at Columbia University and the results demonstrated performance benefits in their approach, suggesting no noticeable performance difference between the operation of the mobile device in a persona compared to the native device operation [Cell02].

2.3   Mobile Separation via Containers

Another approach to providing separation on a mobile device is via Application Containers. This is achieved by using a solution more like Unix method of multiple users where you have one box with multiple users logged in and each user has their own experience. Each user has its own experience and all users run concurrently with one kernel and one operating system [Cree01]. Applying this logic, one user is the personal side and the second user is the enterprise side.

2.3.1   Good Dynamics Technology

Good Technology provides two technologies that address BYOD. First is Good for Enterprise—a secure e-mail, mobile device management and a Intranet-Internet proxy server solution targeted for enterprises. Second is Good Dynamics platform that brings necessary tools, infrastructure, and APIs to developers, allowing them to provide secure applications across devices and operating systems. This protection is delivered by containerizing data at the application level which is accomplished by wrapping a layer of protection around the enterprise deployed apps, which separates the corporate data from the employee's private information and consumer applications. Through this containerized approach, Good Dynamics establishes a secure application environment that minimizes the possibility of data loss. The containerized applications provide the employee the freedom to access enterprise data in a safe and secure manner while being able to switch back and forth with their personal applications without compromising company information. This container-based method applies secure and encrypted transmission of data end to end, from the enterprise servers behind the firewall all the way to the mobile device. Good Dynamics architecture requires that every application use the Good Dynamics APIs and be compiled/linked with its SDK in order for it to run within its secured container. This limits the number of commercial applications that can run within this container.

2.3.2 Divide by Enterproid

Enterproid introduced a mobile virtualization solution in late 2011 called the Divide Platform which gives users a way to use their smartphones for both work and personal life. This solution which runs on Google's Android devices 2.2 or later and Apple's iOS devices like iPhone and iPad is designed to provide multiple profile support, a set of productivity apps, as well as a personal and enterprise cloud management system. The Divide system truly blends in well providing the end user true separation from the work environment without compromising personal freedom [Dolc01].

Divide functions as a container application, with security policies and management features applied solely around that application, leaving the rest of the user's device untouched. Organizations can then in turn manage their own applications within the container. The management features are essentially the same as with most MDM solutions: password policy, encryption, data isolation, clipboard restrictions, remote wipe, screen locking and more [Madd01].

Virtualization is done by having separate secure profiles for work and personal environments. The personal side enjoys the freedom of all the functions, features and applications including access to the Android Market, where as the work side is managed by the IT administrators with a separate more strict profile suited for the appropriate enterprise. Switching between the two environments is done by a simple double tap of the Home key or can also be done via application icons on each of the sides or by going to the notification/alerts bar (Fig. 2.7). Separation of applications is essential in a virtualization system because you want to avoid data leakage from enterprise into personal applications. All applications in the container side

| Personal level of device security to unlock device (I have chosen 4 digit passcode) | Personal side of device Double click home button to go to business side of device | Business level security verification (8 character alpha-numeric passcode) | Business side of device (encrypted data, copy/paste prevention, etc) |

Fig. 2.7 Divide personal/enterprise screen shots

benefit from an encrypted 256 bit storage. Furthermore, encryption is not dependent on the OS, hence not compromised immediately on rooted/jailbroken devices—all the encryption is built within the application. Another example of separation is to limit the potential for leaking enterprise information into personal social networks by limiting the transfer of information between the personal side and enterprise container by restricting the ability to cut/paste information between the two environments. Additionally, Divide provides a rich set of native-Android office applications like mail, calendar, tasks and contacts. The mail application provides threaded email conversations which can also be searched both on the device and on the server. The enterprise mail, calendar, contacts are configured and delivered via the standard Active-Sync API's which provides the ability to connect to various email backend systems like Google, Yahoo, Microsoft, IBM Lotus Notes, etc., All of which are fully synced via 3G or WiFi networks, thereby providing ultimate flexibility wherever you are at. Most importantly, because all these applications run in the separate work environment, everything is fully encrypted and compliant to the policies defined by your IT administrator [Haza01].

Another set of features that furthers capabilities of the system is a cloud based management portal that allows the user and the IT administrator to manage their devices. The user portal, also referred to as "My Divide," provides the user of the smartphone a number of features to control their device, such as the typical device wipe, reset both device and divide password, lock device, but it also provides features like just wiping enterprise data, audio beacon to locate device, device location, push a URL do the device browser and more. The portal also provides additional tabs that give the user the ability to view their network usage, the applications installed as well as a detailed list of the state of the device components like the phone, WiFi, battery, network, audio, location and more.

The IT admin portal, also called "Divide Manager," provides some similar functions such as location, device or enterprise wipe, and password reset, but it also functions where the system is configured by defining groups of users, device policies and enterprise applications. The device policies allow for configuring the typical mobile device management settings, but what is different here is that these settings are for managing the enterprise partition. Here you can set the device password quality, length, expiration, history and lockup timeouts.

Security features allow for controlling clipboard sharing, actions taken when a user removes their SIM card and even checks to see if the device is rooted and gives options on what actions to take if this were to take place. Another key piece of the system is the ability to manage applications in the enterprise partition. The admin is able to upload specific applications that will get pushed to the device upon installation of Divide. Furthermore, the system allows for the ability to control what apps are allowed or not allowed to be installed in the enterprise partition. System performance and battery life are well maintained because these are managed by the operating system and does not require direct access to the hardware [Cree01].

Divide differentiates itself from other solutions is that it runs as an application and it does not require any cooperation with the phone OEM. The install does not require any low-level drivers and uses the standard Android procedures for installing applications. It is a light weight solution that shares much of the device resources and significantly reduces the device overhead required by virtualization and also delivers 256 bit encryption for data [Bran01].

Overall, Divide demonstrates a secure and flexible system for enterprises to use, especially for their BYOD community. It is very functional and provides a wide range of security options and features as well as giving the end user the freedom of their device on the personal side.

2.3.3 TrustDroid

TrustDroid is a practical and lightweight domain isolation solution that runs on the Android OS. It provides application and data isolation by controlling the main communication channels in Android, mainly IPC (Inter-Process Communication), files, databases and, socket connections. This solution is lightweight because it has a low computational overhead and does not require duplication of Android's middleware and kernel like other virtualized solutions. It also organizes applications along with their data into logical parallel domains. Figure 2.8 illustrates different methods for achieving isolation. TrustDroid is shown in Fig. 2.8a where it extends Android's middle ware and kernel with mandatory access control. OS-level virtualization is shown in Fig. 2.8b and this is typically seen in Application level containers. Hypervisor based technologies is shown in Fig. 2.8c. The areas designated in black are the trust computing base (TCB) which is responsible for the security enforcement on the platform and also trusted by the enterprise. TrustDroid has the largest TCB, however, it is one of the most lightweight because it doesn't duplicate any portion of the operating system stack and provides good isolation.

At runtime, all application communications are monitored, as well as access to common shared databases, file-system, networking, and denies any data exchange

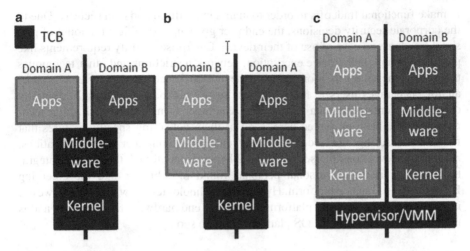

Fig. 2.8 Approaches to isolation [BuDa01]

or application communication between different domains. TrustDroid adds a negligible runtime overhead and compared to other virtualization approaches, only minimally affects performance and battery life [BuDa01].

2.3.4 Android 4.2: Multi-User

Google in its latest version of Android 4.2 OS has the option to have multi-user support which gives the device owner provide separation across users [Cabe01]. This becomes a very interesting development because it gives the device owner the ability to have it's own version of separation—one user for personal use and a second user for enterprise use. It will be interesting to see if enterprises will leverage this functionality in such a way to accommodate and attract the BYOD community to use their personal device for enterprise use as well in this type of environment. In order to achieve this, it will be necessary for the enterprise to deploy its device management software and policies on the enterprise side while allowing the personal freedom on the personal side. This will provide the best experience for the personal side while preserving the security and separation of the enterprise side.

2.4 Summary

The explosion of mobile devices in the consumer space has been quite a disruptor for the mobile enterprise where corporate managed platforms are well defined, well contained and fairly secure. However, now with the emergence of the consumer mobile devices from Apple iOS and Google Android, the enterprise has been forced to make functional tradeoffs in order to maintain platform and data security. Due to the corporate security decisions, the end user gives up a significant amount of personal freedom and ease of use of their device. Enterprise security requirements like password complexity, device encryption, network restrictions and other techniques that restrict the access to information on the mobile device tends to drive users away and/or encourages users to find other less secure alternatives that will eventually compromise enterprise data and access. A number of mobile virtualization technologies have been presented here, each of them delivering specific features that focus on ensuring the security of the enterprise container and applications. Application level containers, type 1 and 2 hypervisors all lack the organic integration of enterprise and personal personas found in solutions like the emerging BlackBerry 10 Balance platform. Hypervisor technologies show promise; however, it is specific to the Android platform, has higher end hardware requirements and is not likely to be seen on the iOS platform anytime soon.

Chapter 3
Mobile Virtualization Comparative Analysis

A way to help analyze which system is most appropriate for your environment is to do a comparative analysis of the various solutions available. Here we compare four solutions: Divide from Enterproid, Horizon Mobile from VMWare, BlackBerry Balance from RIM and Good Dynamics from Good Technology. In order to analyze and understand these various solutions, we have to look at them by each category and the corresponding functions.

3.1 General Platform Support

Each of the solutions provides a Self Service Portal where users can manage and view their device which really helps minimize the end user support costs. Equally important is the management/admin system for the solutions where the IT Administrator can manage the system from any location via a standard browser. The solutions start to differentiate from each other when it comes to what platforms/OS they support. Both BlackBerry and VMWare are platform OS/device specific where as Enterproid and Good Technology support across platform solutions on both Google Android and Apple iOS.

3.2 Device Inventory

The ability to identify what devices your user communities are using reveals a very useful set of statistics for the enterprise. Being able to make decisions based on OS versions or device adoption can save a lot of money and deployment time for applications. All the solutions evaluated do very well at being able to collect device information like device model, manufacturer, carrier, CPU, memory and more. GPS and Location Based Services provide the ability to collect and record device location.

D. Jaramillo et al., *Virtualization Techniques for Mobile Systems,*
Multimedia Systems and Applications, DOI 10.1007/978-3-319-05741-5_3,
© Springer International Publishing Switzerland 2014

3.3 Management Actions

Each of the solutions provide a complete set of device management functions that enable the system to have device or selective wipe, remote lock, device locate, deny email access or user functions like password reset, message sending.

3.4 Security and Policy Management

A key component in each of these systems is the ability to ensure proper device security which is enforced via a set of key device policies. Key security features like requiring device lockup timeouts, minimum password lengths and complexity help ensure that the device is only accessed by the intended person which is also well done by various solutions. Jailbreak/rooting exposes the device to unwanted device access which lends itself to being infected by malware or installing applications that can leak information about the user or the enterprise. Both Enterproid and Good Technology provide very good detection and prevention for this kind of exposure. One piece of functionality that is not available across the surveyed solutions is the capability to interface with other unified device management systems, this being a key feature to help distribute and deploy standard security policies across servers, desktops, tablets and smartphones in an enterprise.

3.5 Enterprise Access

Email, calendar, contacts and instant messaging are key productivity applications for the enterprise, all of which are well supported in each of these mobile device virtualization technologies. As to VPN separation, this is much more complicated based on the virtualization technology. As to VPN access, both BlackBerry and Good Technology provide a built in solution and since both of these provide device level security and policies, they don't have to separate out the network access. In the case of Enterproid and VMWare, VPN has to be filtered at the container level, thereby requiring integration with third party VPN solutions and making the options much more limited for enterprise deployments. The key here is that the data and access to these are separated and protected.

3.6 Application Management

In the enterprise, applications are key to providing information to their users, which make application delivery and management a key component in these solutions. Furthermore, in order to maintain device integrity, access to general application stores needs to be restricted and at the same time, an enterprise application store or delivery mechanism needs to be available. Device policies typically will deliver a core set of applications to the device as defined by the enterprise. Another function that is important is to have the ability to keep an inventory of the applications that are installed on the device per user. The solutions in this space do offer the ability to deliver a core set of applications to the device when installed but they are still very immature in the concept of having an enterprise app store but they do have the foundations for integrating to existing ones.

3.7 Data Leakage Protection

From an enterprise perspective, this is probably one of the more important features/ functions in these systems which are to prevent the leakage of data from the enterprise to the personal side. Many users use their personal email to transfer work related information. Here, a very clear distinction is made between what is work and what is personal. Several restrictions are imposed to help limit the leakage of data. One is by restricting the ability to cut/paste data between personal and work. The second is to prevent attachments in emails from being detached in a non-secure location and using application viewers to securely view the documents. Finally, the more critical function is the ability to attach or associate an application with the container such that all data stored by that application is placed within the secure container. Both Enterproid and Good Technology have the ability to integrate an application into the container, however, Good Technology requires that the application be recompiled with their special SDK libraries, whereas Enterproid has a simple binding process where you can upload the application and it will add specific hooks to the executable that will associate it with the container.

3.8 Comparative Results

The BYOD phenomenon is real and it is affecting every enterprise. Some allow those users on their corporate network with enterprise policies but users only remain on for a short time because they don't want to give up their personal flexibility. They find corporate policies too restrictive and do not want to lose too much personal freedom. As such, enterprises are losing many users and wasting valuable resources in the process. They are quickly realizing that they need to invest in device virtualization technologies that will allow them to maintain their security integrity as well as keep the personal freedom and flexibility of the user because it is becoming very clear from various BYOD surveys that employees are willing to pay for their device and usage and enterprises are willing to provide the necessary infrastructure for them to access their networks. Due to this, many enterprises are evaluating solutions such as the ones discussed here (Table 3.1).

Table 3.1 Mobile virtualization comparative analysis

Category	Functionality	Enterproid Divide	Horizon VMWare	BlackBerry Balance	Good Technology
General platform support	Self service portal/management admin portal	✓	✓	✓	✓
	iOS—iPhone, iPad, iPod touch support	✓	✗	✗	✓
	Android 2.2 + smartphone and tablet support	✓	✓	✗	✓
	BlackBerry OS support	✗	✗	✓	✗
Device inventory	Model, manufacturer, carrier, name, user	✓	✓	✓	✓
	Memory, external storage, battery info, serial #	✓	✓	✓	✓
	Device location (GPS)	✓	✓	✓	✓
Management actions	Device wipe, selective wipe, remote lock	✓	✓	✓	✓
	Reset pwd, send user msg, deny email access	✓	✓	✓	✓
Security and policy management	Password policies—8 alphanumeric password	✓	✓	✓	✓
	Device lockup—multiple timeout support	✓	✓	✓	✓
	Jailbreak/root detection/actions	✓	✗	✗	✓
	Unified Mgmt, incl. traditional endpoints	✗	✗	✗	✗
Enterprise access	Email, calendar, contacts, instant messaging	✓	✓	✓	✓
	VPN separation/isolation	✗ᵃ	✗ᵃ	✓	✓
Application management	Prompt user to install required apps	✓	✓	✓	?
	Prevent access to external app store	✓	✓	✗	?
	Recommend apps/enterprise app store	✗	✗	✓	✓
	View installed apps	✓	✓	✓	?
Data leakage protection	Prevent cut/paste from enterprise/personal	✓	✓	✓	✓
	Prevent/manage email attachment export	✓	✗	✗	✓
	Enterprise app store container integration	✓	✗	✗	✓

ᵃSome support available

By far the most well managed system is that of BlackBerry Balance because it provides proven device level security and data integrity as well as very good integration with the device and device management system. Compared to the other systems, it is restrictive due to the device lockup policy being device wide. The BlackBerry Balance system would be much better for personal use if it provided a way to containerize all the enterprise apps and data such that only that section would require the device lockup and timeout. Everything else would be fine because any wipes would be that of only enterprise data and applications, thereby leaving the personal information intact. The one obvious downside to this solution is that it is BlackBerry centric and it only works for BlackBerry devices. Given the large trend and influx of iPhone and Android devices, this solution is limited and does not solve these problems for the range of diverse devices that users have.

As to device virtualization via the use of a hypervisor, VMWare really has something very interesting and powerful with its VMware Horizon Mobile platform. It provides the most flexibility and gives the ultimate separation; however, it has some limiting factors that prevent it from wide adoption at the moment. One limitation is that the solution requires that the device manufacturers incorporate the hypervisor into the device image for security purposes and on top of that, enterprises need the ability to ensure that the hypervisor has not been compromised in any way. As such, some types of security applications like malware detection software and device management agents will need to reside on the personal side in order to ensure that the device has not been compromised.

Lastly, Divide from Enterproid provides a very attractive solution for the user community because it provides the best flexibility out of the three solutions. The product is really addressing the sweet spot in the market by addressing most of the security requirements as well as addressing the personal requirements from the employees. Many enterprises are looking at this solution because it provides the flexibility to incorporate their own corporate policies, enterprise applications and device management.

Chapter 4
Mobile Virtualization Case Study

In order to see how mobile virtualization affects the BYOD user, a pilot was done with Divide from Enterproid, one of the mobile virtualization technologies described previously that is based on a application container approach. The pilot was run for 12 weeks with over 1,100 registered users out of which 817 were activated. Users were all employees who volunteered to participate in the pilot. There were no screening criteria and no specific instructions for use of Divide during the pilot. There were no prescribed evaluation tasks; participants were free to use Divide on their personal smartphone and/or tablet device(s) for business and personal tasks, in a manner natural to them.

4.1 Pilot Start

Prior to starting the pilot, several months were spent going through an evaluation process with Enterproid to work through all the requirements to launch the pilot. These requirements included testing the various models of Android and iOS devices and applications along with their configurations. This setup was tested with ten users and once we reached a point where we reached out pilot entry criteria, we then launched the pilot. The pilot entry criteria consisting of ensuring that there were no major outstanding installation, security and usability issues. Initially we opened up the pilot for general registration and users rapidly signed up for the pilot. In the first week, 80 users were enabled in order to have a controlled start which allowed for making any adjustments in the deployment process before adding more users. We made a few updates were made to the documentation and then proceeded to board more users. Starting the second week and for 2 more weeks after that, an additional 200+ users were boarded each week. On the fifth week we boarded approximately another 100 users and after that we started slowing down because we were reaching the 800 user limit for the pilot. Also, 4 weeks into the pilot, it was decided that the pilot would only be run for a total of 12 weeks, and as a result, the number of active

D. Jaramillo et al., *Virtualization Techniques for Mobile Systems*,
Multimedia Systems and Applications, DOI 10.1007/978-3-319-05741-5_4,
© Springer International Publishing Switzerland 2014

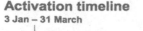

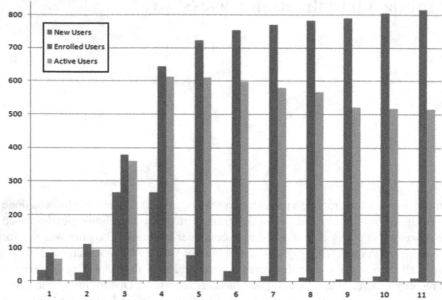

Fig. 4.1 User activations

users started to decline because a number of people noted that they did not want to continue using a technology that was not going to be deployed for general use. Overall, for the duration of the pilot, out of 800+ registered users, 500+ users were active, indicating a healthy adoption of the technology from the users and continued using it on a regular basis (Fig. 4.1).

4.2   User/Device Comparison

The first 3 weeks of the pilot, we refined the environment and process in preparation to the full scale deployment. During this time, registrations (signups) were open but activation was not fully initiated until the fourth week. In those initial setup weeks users were selectively enrolled in order to verify the end to end process. On the fourth week, enrollment was completed for all users registered for the pilot. In Fig. 4.2, the red and blue bars represent the number of enrolled users each week. Weeks 4 and 5 showed a significant spike in activation due to increased interest in the pilot. The green bar represents the total number of devices on which the container technology had been installed and activated. Note that the red and blue numbers do not sum equally to the green; this is because many users activated more than one device. The breakdown of users and device coverage for the pilot is explained in the next section below (Fig. 4.3).

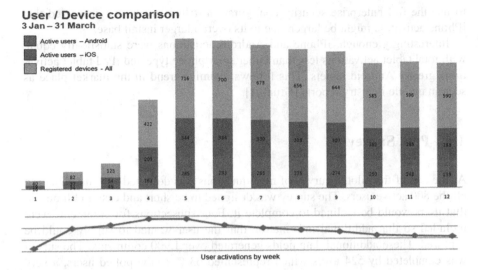

Fig. 4.2   User/device comparison

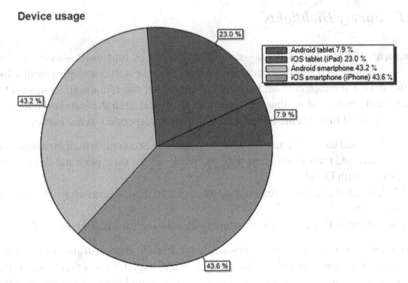

Fig. 4.3   Device distribution

4.3   Device Usage Distribution

One interesting fact is that as the pilot was started, it was not known what the device distribution would be. We speculated there would be many users with tablets because that was mostly their own personal device and they did not want to commit to use the full enterprise security configuration. Alternatively, we suspected that iPhone activation might be larger, due to its overall larger install base.

Interestingly enough, iPhone and Android activations were statistically equal, with total tablet activations less than either smartphone type and iPad tablet activations greater Android tablets. This follows a similar trend in the market place as seen in mobile industry reports [Grum01].

4.4 Pilot Survey

At the end of the pilot, a survey of nine questions was designed and distributed to all the 800 active users. The survey was designed to be short and easy to fill out, so that users would be inclined to complete it. Four subsections for comments were included so that additional information that the user wanted to supply could be captured. These additional data fields generated over 1,900 comments. The survey was completed by 524 users which represented 63 % of the polled users, a very successful response that exceeded our expectations.

4.4.1 Survey Highlights

In summary, 74 % of pilot users—roughly three out of four users—said they preferred a containerized solution compared to the general enterprise deployment which has full device management and security controls. Of the remaining sample, 11 % were neutral, preferring neither solution, and 15 % preferred the standard enterprise solution. Several findings are summarized from that responded to the survey:

- 80 % indicated they could use Divide to successfully accomplish their business tasks.
- 75 % indicated their interaction with the work side of their personal devices was improved with Divide.
- 64 % indicated that they preferred a more relaxed device level security password requirement.
- 84 % indicated that they used the container solution on a daily basis (78 %).

At a high level, the pilot was very successful and it demonstrated that the user population is very interested in this kind of technology for the enterprise. Survey results showed that 86 % of the users would recommend the technology to fellow colleagues and 98 % of the users requested that they be notified of future pilots.

4.4.2 Survey Analysis

The survey, as mentioned previously, consisted of nine questions with each question focusing on a different user experience attribute: application performance, overall satisfaction, usability, capability and usage.

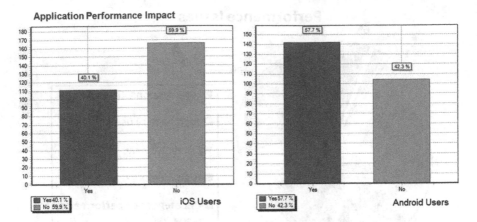

Fig. 4.4 Application performance impact

4.4.2.1 Application Performance

Application performance was a qualitative observation as to how the system responded when the device was used by the users. Users reported that they based their response on how quickly the applications started up, response time of the applications and the corresponding data that the applications provided. During the course of the pilot, we noticed that users had a degraded performance experience on Android devices compared to that of iOS devices. Therefore, as part of the survey, the type of device being used was recorded so that the difference in performance across the platforms could be measured.

As indicated in Fig. 4.4, 40 % of the iOS users compared to almost 58 % of users on Android indicated that they noticed a significant degradation in performance compared to the non-virtualization environment. This may seem like high numbers, but from the survey results, in general, this was an issue that really only affected power users.

Performance issues were furthered categorized by the type of issue as shown in Fig. 4.5. Each of these issues was identified early on in the pilot; as such, these questions were included in the survey to help determine the pervasiveness of each of these issues. Each of these issues is explained as follow:

- High battery usage was the most prevalent issue, accounting for 62 % of the users reported performance issues. After investigating this result, it was determined that this can be attributed to two factors:

  - Users in general were using the piloted solution more than the regular solution.
  - Email sync on the piloted solution does not have the ability to schedule sync times accounting for the higher battery consumption.

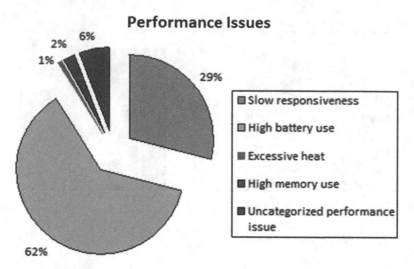

Fig. 4.5 Performance issues

- Slow responsiveness of the solution and/or device was the second most prevalent issue, accounting for 29 % of the reported performance issues. After investigating this result, it was determined that this was attributed to three factors:

    - The solution has a large memory footprint due to all the overlapping functions that it needs to provide. These functions are already part of the operating system but need to be containerized/separated.
    - Investigating with the solution provider, it was found that on Android, several of the functions were being done on the main application thread which caused the application being non-responsive making application tasks take more than a few seconds. Later on in the pilot, an update was later provided that remedied the issue.
    - A number of the devices in question were of average to low memory and CPU configurations. Once this solution was installed, the whole system was stressed with low memory conditions, making the system very sluggish as reported by the users.

- Excessive heat accounted for 1 % of reported performance issue; this was found to be normal due to the high intensity in the mail replication, higher CPU and battery usage.
- High Memory usage accounted for 2 % of reported performance issues which accounts for those power users, mostly on Android that have the knowledge and ability to look at the memory consumption of applications.
- 6 % of the reported performance issues were not categorized. This represents performance problems that users noticed and reported, but with insufficient detail to be able to assign to specific categories as above.

4.4.2.2   Overall Satisfaction

It is very evident from this metric that regardless of the performance, battery, memory usage issues, the concept was very well received. 42 % of respondents indicated they were very satisfied and 38 % indicated they were somewhat satisfied yielding an overall positive satisfaction rating (sum of very satisfied and somewhat satisfied) of over 80 %. This gives an overall indication that this type of solution is something that BYOD users would be very interested in (Fig. 4.6).

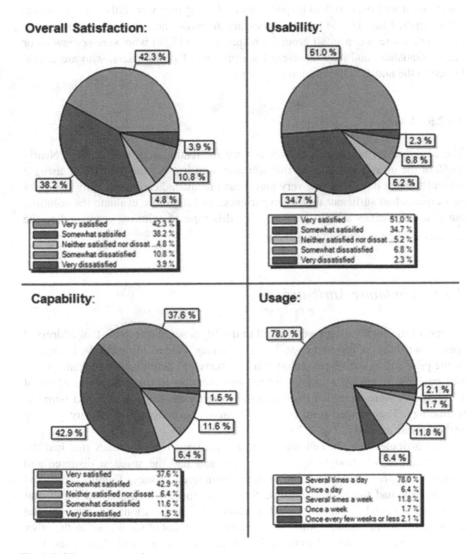

Fig. 4.6 Pilot survey metrics

4.4.2.3 Usability

The results show that the solution has a high usability rating, with 86 % of respondents indicating they were either very satisfied or somewhat satisfied with the ease of use of the solution. This is also a very positive indicator (Fig. 4.6).

4.4.2.4 Capability

The results show that users were fairly satisfied with the capabilities provided with the solution with over 80 % of respondents indicating they were either very or somewhat satisfied that Divide has the necessary features and functions to perform as expected. There was a small group of respondents (13 %) who were somewhat or very dissatisfied, and may represent the opinions of power users, who are accustomed to the non-BYOD solution (Fig. 4.6).

4.4.2.5 Usage

The response to the usage question is very interesting and encouraging. Nearly 85 % of the users who installed this solution used it every day with 78 % using it several times a day. This is a very significant result indicating not only that pilot participants had sufficient time to experience, and therefore evaluate the solution, but that there exists a high user need for this type of solution, compared to the existing enterprise offering (Fig. 4.6).

4.4.3 Container Attributes

As part of the survey, users were asked to qualify a set of questions that addressed specific attributes of the container. These attributes were identified at the beginning of the pilot and as such were included in the survey to help identify the pain points and benefits of having a mobile virtualized solution in the enterprise. A set of 15 attributes were identified to isolate the experiences that the user had with the technology. They ranged from installation, application usability, security, privacy and others as defined in Fig. 4.7 below.

The attributes that showed the greatest sensitivity were the ones that had the highest values for strongly agree and agree and had the smallest disagree and strongly disagree. Analyzing the results, Structure, Privacy, Finding apps and Productivity had the highest ratings. Second to these which showed significant importance were Installation, Email, Consistency, Applications, Interaction, and Comparison to the standard solution. On average, but not as significant as the other attributes, were Calendar, Contacts, Notifications, Settings and Passwords. One noteworthy observation from these results is that Password is no longer a big issue.

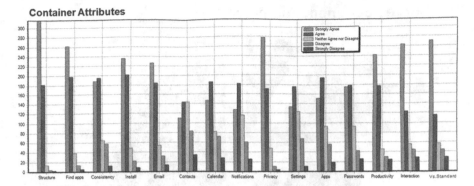

Fig. 4.7 Mobile virtualization container attributes

This is likely because with Divide, the strong password requirement from the device side is moved to the virtualization container. Users understand that security is important, but that it should be relevant to the data that is being protected.

4.4.4 Pilot Survey Comments

As a final way of capturing the results from the pilot, it was determined that the users have the greatest amount of flexibility in reporting back their experience with the technology being piloted.

As seen in Fig. 4.8, 51 % of comments users submitted were about the overall performance of the solution which is also shown in Fig. 4.5. The second most frequently reported category of issues related to core features in the system. These included Calendar, Contacts, Email and Phone. Application management was the next largest category of issues, which corresponds closely to the comparative analysis previously done in Chap. 3. These solutions do not lend themselves to easy application management due to the corresponding security requirements. In order to provide a better user experience, the solution needs to be able to integrate with the Enterprise App Store so that application updates can be delivered in a more timely fashion.

4.5 Pilot Summary

The pilot showed that the Container technology was well received by enterprise BYOD mobile users. Users felt that it liberated their personal device; it allowed them to use their device anywhere and anytime without any restrictions. As

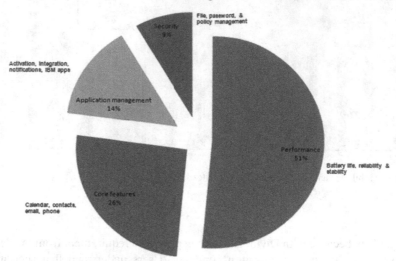

Fig. 4.8 Pilot survey comments summary

testimony to this, many users did not want to go back to the standard enterprise offering and in a follow-up survey it was discovered that they did not return to the enterprise offering. However, there were a number of challenges and improvements are definitely required. Needed improvements include greater battery life and sync performance, improved integration of contacts across the personal and enterprise side, improved email/calendar client application and enhanced synchronization options. From a deployment perspective, the deployment of the container technology needs improvements from the administration and management console side in order to have a better boarding process. Along with this, application management in the container needs to be improved from the deployment perspective, so that more applications can be deployed more easily into the container as well as the update management for these applications. Essentially, the technology needs to provide the ability to integrate the ability to deliver and update applications with the enterprise app store. Another key component that was lacking at the time of the pilot was the integration of the container device management system with the enterprise device management system. In conclusion, the technology was adopted very quickly and very well received. Even though the functionality still needs more maturity, it shows great promise for the BYOD community and to the enterprise.

Chapter 5
Mobile Virtualization Reference Architecture

An agile, lightweight, but secure extensible hybrid application container and deployment mechanism for mobile devices has the potential to provide improved cross platform support, reduced development lifecycle time and a consistent application security model within large organizations. Such an architecture can provide a set of security and mobile device management interfaces to lightweight applications written in a high level markup and scripting language and also how these applications can be provisioned via push or pull mechanisms to an enterprise user's device. The use of such a container for hybrid applications widens the potential support and development resource within an enterprise possessing readily available skill sets, thus permitting mission critical applications to be developed in a much shorter time frame [JaSm01].

5.1 Mobile Virtualization Application Container

Mobile Virtualization Application Container is a hybrid application container model that builds upon the ideas of solutions such as Apache Cordova (PhoneGap) [Phon01] but extends this idea such that the container shell application becomes home to not just one single use application but multiple applications. These are presented to the user as an extensible virtual desktop within one application running on the device. Additional hybrid mobile applications can be dynamically downloaded and installed within, their data and services managed and encapsulated within the application space of the container.

This hybrid mobile application (HMA) is created to run on a mobile device where the application logic is primarily written in JavaScript and the user interface components are defined using HTML5 and CSS3.

The HTML and associated JavaScript application logic are contained within a native code wrapper which allows what would otherwise be a web application to be installed as an application on the mobile devices desktop. The native wrapper exposes

D. Jaramillo et al., *Virtualization Techniques for Mobile Systems*,
Multimedia Systems and Applications, DOI 10.1007/978-3-319-05741-5_5,

device level function to the scripted portion of the application through a set of plug-ins written in the native language of the particular mobile platform, each plug-in has an associated JavaScript API. The plug-ins can expose native function that would otherwise be inaccessible using standard HTML5, for example access to the device calendar, phonebook or through built in hardware such as the camera.

The interaction between the JavaScript API and the native code is implemented on each mobile platform using methods made available through the platform SDK. Calls from JavaScript to native code exploit the ability to intercept navigation events and examine the URI scheme, a URI scheme can be used to indicate that the URI is an encoded method call and should not be passed on to the embedded browser to handle, the hybrid application shell can then process the encoded URI and call native methods within the application shell. Conversely, function calls can be made to the embedded browser code by injecting JavaScript at runtime into the currently loaded web page.

5.2 Benefits of Hybrid Application Development

There are several advantages to using a hybrid development approach when creating mobile applications. Using HTML5 and JavaScript increases the portability of the application; the UI and application logic can be reused across multiple platforms with minor changes to presentation format performed using CSS [AlLu01]. Another major advantage is the abundance of HTML and JavaScript development resources in comparison to native language knowledge of Objective-C and Java.

5.3 Hybrid Application Container Requirements

Increasingly large organizations are allowing employees to use their own mobile devices for work purposes; this has created a set of Mobile Device Management (MDM) challenges for the IT teams within these organizations. A major challenge is to allow the employee freedom to use the phone for personal tasks unencumbered by lengthy passwords and IT Management enforced policy, yet still protects enterprise data and applications. In order to do this some type of separation mechanism needs to be introduced in order to ensure that there are no data leakages or security intrusions. Protection is delivered in the form of a virtual container that applications can run inside, their data is containerizing at the application level, this is accomplished by wrapping a layer of protection around the enterprise deployed apps, which separates the corporate data from the employee's private information and consumer applications. Further security is provided by some solutions by using specific SDKs that intercept all data related APIs and store the data in secure/ encrypted file stores. An additional security measure is to monitor the application using Mobile Device Management and provide functions that give an IT administrator the power to do corporate wipes of the enterprise container contents. A time bomb mechanism may be employed that will self wipe the container and data if the device has not checked in within a period of time defined by a device security policy. Other solutions like micro-containers have been done which use cloud based services to provide mobile service-containers for hosting service-based applications [OmSa01].

A few steps/requirements are needed in order to create a functioning Hybrid Application Containers which include Hybrid Application Creation, Application Deployment and Update Management, Mobile App Store, Hybrid Container API and Hybrid Container Security.

5.3.1 Hybrid Application Creation

Creation of a hybrid application should not require a developer to possess platform specific skills but rather have a grasp of common web development methods including knowledge in HTML5, CSS3 and JavaScript. The container itself consists of a native application code shell and provides all the necessary services and APIs that an application team would need to build applications. This allows the creators of hybrid applications to focus on the presentation and business logic in a platform independent manner without having to learn new high-level languages such as Objective-C or Java. Web development skills are more widely available in an organization; as a result the pool of developers who can contribute custom applications will in most cases be substantially larger than the set of developers who possess knowledge of the high level languages used to create applications on platforms such as iOS and Android.

5.3.2 Application Deployment and Updates

Application deployment as well as application updates are essential in a container environment and they need to be simple and secure. Application deployment can be performed in two ways. The first by pushing the application(s) during initial installation of the hybrid container, the second deployment is via user requested installation from an app store. The configuration of the applications to be installed is controlled by the backend system that manages or communicates with the hybrid container. In the case of application updates, there are two types that applications are sensitive to. One is the core native shell where plug-ins reside and the second is the internal HTML/CSS code. Depending on the complexity of the application, there may be few updates to the shell and frequent changes to the UI. In the case of updating the shell, this requires a complete re-installation of the application including both the shell plug-in as well as the core UI files. Careful consideration needs to be taken to make sure that any files or settings from previous installations do not conflict or cause adverse effects with the new installation.

5.3.3 App Store Model

Hybrid applications by nature are very dynamic and are continuously being updated, new applications may be released that users will want to install after the initial setup of the hybrid container. In order to address this scenario, the hybrid container can be architected to connect to an application store backend that will allow the user to select from a catalog of applications that they are entitled to download and install.

The enterprise can also have a managed set of required applications that would be pushed to the hybrid container ensuring that users always have a set of pre-defined applications.

5.3.4 Hybrid Container API

The container itself must be written in the native language used by each targeted mobile platform. It will provide a consistent set of APIs to application developers; it will also contain the mechanisms to allow hybrid applications to be installed dynamically within the container.

5.3.4.1 Security and Storage

Security is of utmost importance within a large organization, data may have varying levels of sensitivity but the determination of what is sensitive should not be left purely up to the individual application developer. The best policy for application development is therefore to provide secure storage through strong encryption for all application data. Allowing the container to handle this behind a simple storage API reduces the burden on the application developer and allows IT management to be confident that standards and policy are being adhered to.

Due to the nature of the hybrid container application, data separation must be enforced via strict name spacing to prevent one application reading another's data. It may however be useful to allow sharing of data in certain circumstances e.g. allowing an instant messaging application access to the database of a contacts application.

5.3.4.2 Enterprise Network Access

Providing access to data and services that are located within an organizations internal network from a mobile device can be problematic and a source of much debate amongst security teams. The option to provide device level Virtual Private Network (VPN) exists but could expose the organization to attacks by malware that may have been unintentionally installed on an employee's personal device. A second option is to provide access on an application-by-application basis often through a secure reverse proxy mechanism. This is much more secure but can be a lot of effort for the IT management part of the organization. The ideal option is to allow the container to manage access providing a common interface for applications that reside within it.

5.3.4.3 Notifications

Applications in the most part act in a request response manner; however there are occasions where it is necessary to push information to an application. In a mobile context this usually takes the form of a push notification sent from a messaging provider service usually run by the creator of that particular mobile operating system. Apple provides the Apple Push Notification Service (APNS) whilst Google provide the Google Cloud Messaging (GCM) service. It is necessary to abstract the use of these messaging providers away from application developers both for client and server side development [JaNe01].

5.3.4.4 Email/Calendar/Contacts

Common applications such as email, calendar and contacts could be provided with the container rather than on an organization-by-organization basis. This function is required to reside within the container as the information used is of a highly sensitive nature. The protocols used for these applications are well documented and standardized so implementing them as common components is justifiable.

5.3.4.5 Application Updates

An update mechanism must be present in the container logic providing two distinct functions. Firstly the ability to check for code updates to the container itself, these may be frequent as the demand for new native plug-in components is driven by the arrival of different applications. Secondly to check for new versions of previously installed applications within the container. Both methods of update will require communication with a server based management component.

5.3.5 Hybrid Container Security

The container should provide an authentication mechanism that secures access to the applications and data residing within. The credentials used by the container will be determined at install time and may be governed by a policy which enforces a password change at a time interval that is configurable.

5.4 Hybrid Application Container Creation

The decision was made to use the open source Apache Cordova project rather than create the hybrid framework of the container from scratch, Cordova already has a clean extensible plug-in architecture and a large community of developers creating plug-ins to offer additional functionality on top of the base set.

When a Cordova application starts up the native shell code loads the HTML/ JavaScript UI application logic from a fixed www root folder within the application file system. This root folder is read only once the native application is installed so it is necessary to subclass the Cordova class that loads from the root folder, this allows the UI application logic to be loaded from a different writable folder that we can install downloaded hybrid applications within.

5.4.1 System Architecture

Figure 5.1 shows a typical deployment configuration for the container architecture and required backend services. A mobile client running the container connects over the Internet to a secure reverse proxy that can perform device authentication and

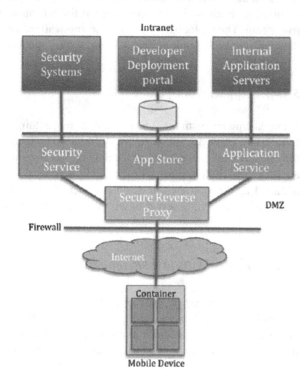

Fig. 5.1 Hybrid container deployment architecture

authorization to establish a secure session. Within this secure session the device can contact a variety of services. These services are located behind a firewall in a demilitarized zone (DMZ).

The App Store service allows a user to browse and download new applications to the container; it also provides version management services to the container. Individual applications will need specific backend services made available to them, these can all be accessed via an application service located inside the DMZ. Requests to the service will be distributed to many pre-existing service APIs located within the organizations intranet.

A Security service also resides within the DMZ which provides applications with the service APIs that allow various authentication and authorization tasks to be performed. This service layer also accesses a variety of pre-existing intranet based applications. These can include employee directory servers, Mobile Device Management (MDM) solutions or authentication services. Figure 5.1 demonstrates system architecture for a hybrid container deployment.

5.4.2   Hybrid Container Components

The Hybrid Container is made up of the following components: Container Application, Plug-ins that handle security, device hardware features, messaging and storage; and Application Manager.

5.4.2.1   Container Application

As illustrated in Fig. 5.2 the container application is made from a variety of layered components, the application itself is written in the native language of the mobile device i.e. Objective-C for iOS, Java for Android. The container contains bootstrap code to load the initial Application Desktop user interface web application in a Web View when the application starts. An application executing in the Web View can call native code in the form of plug-ins, each plug-in has a JavaScript API and a corresponding native language class implementation. The desktop contains icons which when selected by the user trigger a call to the Application Manager (AM) which then loads the selected application into the Web View components.

5.4.2.2   Plug-ins

A base set of plug-ins are included with the container, however more may be added over time as the container receives updates. Plug-ins are written to conform to the Apache Cordova interface and many of the standard Cordova plug-ins are included. Each plug-in has a corresponding JavaScript API so that it may be called from a hybrid application.

Fig. 5.2 Hybrid mobile
application container

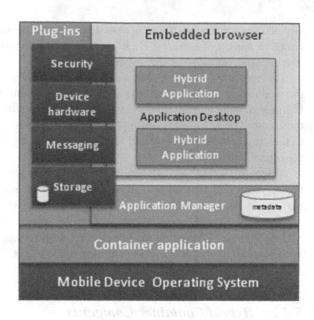

Security

This set of plug-ins provides authentication and authorization components used when accessing services hosted on the organization's intranet, this includes a filtered VPN service.

Device Hardware

Access to hardware-based functions of the device such as GPS, Camera and Microphone can be obtained and interacted with through this set of APIs.

Messaging

The messaging plug-in set has responsibility for processing incoming notifications and ensures that the message is routed to the correct hybrid application. The AM is called if the application is not currently running and an appropriate alert can be shown allowing the user to switch to the application that the push notification targeted.

Storage

In order to maintain security standards with regards to storage, any write operations performed by a hybrid application should go through the storage API. This allows the data to be encrypted before it is written. If the application requires access to a database then an application specific SQLlite instance is created. The entire database is encrypted using the container password created when the container is first installed.

5.4.2.3 Application Manager

The Application Manager shown in Fig. 5.2 has several responsibilities including loading the desktop on the first instantiation of the container. When applications other than the desktop are started the container stores the state of the current application then reads the new applications descriptor. Any plugins required by the application are instantiated and the web resources loaded into the web view component, if a previous state was stored the state is reloaded before the application displays. The AM handles the installation of new applications within the container and also periodic version checks for both installed applications and the container itself. Version checks are performed by making an HTTP request to a server based version component within the app store, applications can be updated dynamically and restarted within the container. In order to update the container itself the full application must be reinstalled.

Once an application has been developed the application resource files are added to an archive file and zip compressed. Along with the resources in the archive are an application descriptor and a desktop icon used to display the application within the container.

The application descriptor contains metadata such as the application name, version, the plug-ins used by the application. Installed application details are maintained within a local database by the container and queried at various points in the applications execution cycle.

5.5 Mobile Virtualization Container Interface

In order to integrate within the enterprise, the Mobile Virtualization Container needs to integrate with the Enterprise Mobile Device Management and Bring-Your-Own-Device Management systems together using Mobile Virtualization Container Interface (MVCI) for mobile devices. This method integrates the various security components of the mobile device management system into a layered architecture that allows the enterprise to deploy and maintain their device policies onto one or more mobile virtualization systems.

One of the challenges in an enterprise is how to perform device management for the entire company. Servers and desktops have their own systems and now Mobile has their own as well. This increases the number of device management systems because none of Server/Desktop and Mobile systems integrate with each other. In order to provide the most optimal systems management, these systems need to integrate with each other into a centralized system and at same time they need to take advantage of device management capabilities available in each of the platforms.

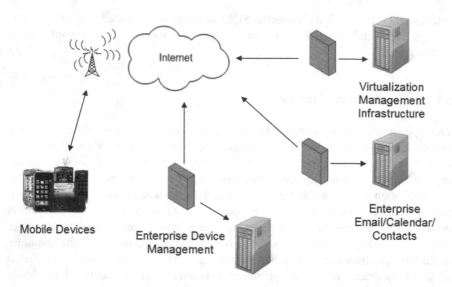

Fig. 5.3 Example of enterprise mobile virtualization system

Figure 5.3 shows what a typical Enterprise Architecture would look like when it supports a Virtualized Mobile Container. This architecture needs to integrate with the existing Enterprise Device Management system and also with the Virtual Management Infrastructure that supports the Virtual Mobile Container on the device.

5.5.1 MVC Interface Specification

Through user scenarios we can analyze the various types of use cases that a mobile virtualization system needs to support. One scenario could be where a user may receive an SMS message from their work where the system detects the event and determines that it is coming from a work related number. At that moment, the system immediately changes the user profile to the enterprise profile and sets the appropriate polices and security profiles upon entering that virtualized environment. Another scenario could be a location based such that when the user enters the enterprise location where the enterprise WiFi is detected at which point, the secured WiFi network and browser policies are engaged and the enterprise container is then ready to be used by the user.

5.5.2 MVC Architecture

In order to achieve this level of integration, a system called the Mobile Virtualization Container Interface (MVCI) needs to be devised such that it interfaces with the enterprise device management system and also to the virtualized device management system.

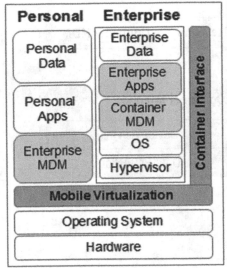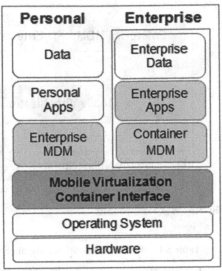

Fig. 5.4 MVCI in mobile virtualization systems

The interface should be able to push and query device management policies depending on the user profile being applied due to location or by user's choice. The importance here is that the Enterprise Device Management agent can reside outside the container so that it can monitor the device for any changes in the system like installing enterprise applications outside of the virtual container or leaking data outside the virtualized container.

In Fig. 5.4, MVCI demonstrates how it is integrated in both a Hypervisor and Container solutions. MVCI provides interfaces for setting, getting and notifications for its components which are described in the next section. The key item to note here is that there is application flow that is introduced here that allows the Enterprise MDM to both query and set policies to the Container MDM as well as being able to register for notification for any specified events.

5.5.3 MVC Components

The MVCI is made up of a set of components that provide a set of services to and from the Enterprise MDM and the Container MDM (Fig. 5.5). These services provide functions to query device information like physical location, device attributes and more. For this level of control, we need to decompose the access down into four components as described in Table 5.1.

Each of these components—Device Query, Device Actions, Policy Management and Data Protection all play a specific role in managing and controlling the device container and interfacing with the Enterprise MDM. In order for the Enterprise MDM

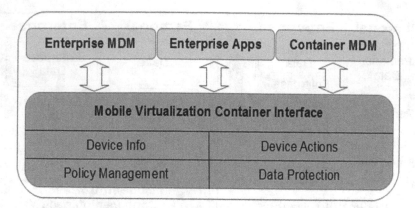

Fig. 5.5 MVCI components

Table 5.1 Mobile virtualization components

Components	Functionality
Device query	Model, manufacturer, carrier, name, user
	Memory, external storage, battery info
	Device location (GPS), IMIE, serial #
Device actions	Device registration, selective wipe, remote lock
	Reset password, send user messages, deny email access
Policy management	Password policies—length, age, complexity
	Lockup timeouts
	Jailbreak/root detection/actions
	Device restrictions (cloud backups, etc.)
Data protection	Prevent cut/paste from enterprise/personal
	Prevent/manage email attachment export/preview
	Enterprise app store container integration

(EMDM), Enterprise Apps (EAPPS) and the Container MDM (CMDM) to be synchronized, the Mobile Virtualization Container Interface (MVCI) provides the necessary interfaces for the two systems to communicate with each other—see Fig. 5.4.

Each of the MVCI Components—Device Query, Device Actions, Policy Management and Data Protection each provide an integral part of system and are described as follows:

5.5.4   Device Query

This component is responsible for collecting device information like device identification information; device characteristics like memory, storage, camera, sensors, location, IMIE, serial #, etc. so that it can be made available to the registered Enterprise Mobile Device Management Systems.

5.5.5 Device Actions

This component is responsible for executing device actions received from the device management system such as device wipe, partial wipe, remote lock, reset password, send a message to the device and block the receiving of emails.

5.5.6 Policy Management

This component is responsible for enforcing device policies on device screen lock password length, strength, quality, expiration, history and timeout. Other policy management features would include controlling the storage of data in the cloud and also the detection of device rooting/jailbreaking.

5.5.7 Data Protection

This component is responsible for controlling and ensuring that data is maintained within the container. Such actions include cutting and pasting to and from the container, and managing the viewing of email attachments, which is a critical component in preventing data leakage. Another important part of this component is the managing of the applications that can run within the container.

5.6 MVC Interface System Design

In order to demonstrate the functionality of MVCI, a set of use cases demonstrating the interaction between the enterprise device management system, the application and the container, show how each of the components are integrated.

5.6.1 Device Query

This component being responsible for collecting device information is broken down into a series of functions to support the delivery of device information. Figure 5.6 details the sequence diagram for the functions supported by the Device Query component and are described as follows:

- Device id is a unique id that is used to identify the device for applications that want a way to identify the device in a way that can be differentiated from other devices. Enterprise MDM makes the request to MCVI and in turn requests the id

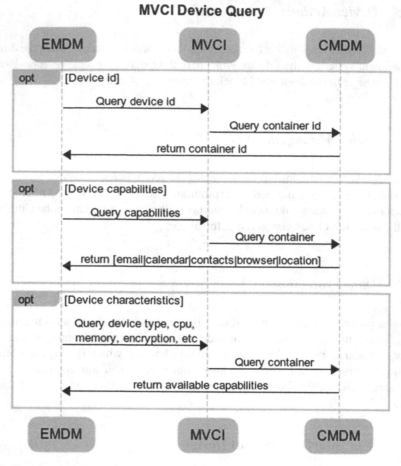

Fig. 5.6 MVCI device query

from the Container MDM which is usually a unique id by the container who what a unique container device id.

- Device capabilities are used to identify the software functionality like email, contacts, browser and location. As shown in Fig. 5.6, Enterprise MDM makes the request to MCVI which in turn requests the capabilities from the Container MDM and is then returned back to EMDM.
- Device capabilities are used to identify the software functionality like email, contacts, browser and location. As shown in Fig. 5.6, Enterprise MDM makes the request to MCVI which in turn requests the capabilities from the Container MDM and is then returned back to EMDM.
- Device characteristics are used to identify the devices hardware functionality like CPU, memory, encryption, screen, touch, Bluetooth. As shown in Fig. 5.6, Enterprise MDM makes the request to MCVI which in turn requests the capabilities from the Container MDM and is then returned back to EMDM.

Fig. 5.8 MVCI policy
management

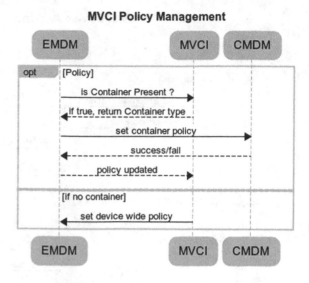

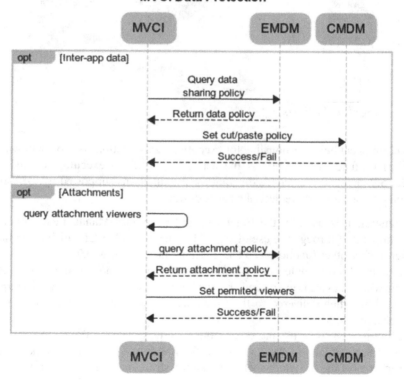

Fig. 5.9 MVCI data protection

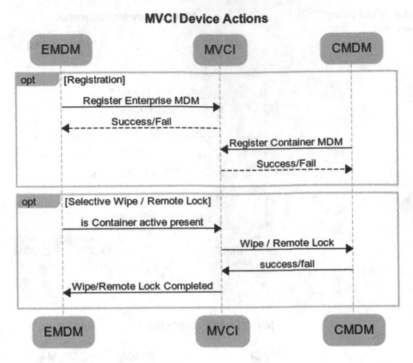

Fig. 5.7 MVCI device actions

5.6.2 Device Actions

This component being responsible for executing device actions is broken down into a series of functions to support the actions required to execute the functions. Figure 5.7 details the sequence diagram for the functions supported by the Device Action component and are described as follows:

- Registration is used by the Enterprise and Container Management system to identify itself through the use of signed certificates to MVCI which is the initial step before other functions can take place as shown in Fig. 5.7.
- Selective Wipe/Remote Lock is a function that an admin/user function typically uses when a device is misplaced or lost. Command is initiated from the Enterprise MDM and then delivered to the Container Management system via MVCI as shown in Fig. 5.7.
- Policy is used to verify and deliver the Enterprise Device Policy. As shown in Fig. 5.8, this function checks to see if the policy is up to date sets the security properties of the container such as password strength, length, expiration.

5.6.3 Data Protection

This component is responsible for controlling and ensuring that data is maintained within the container. Such actions include cut/paste to and from the container, managing the viewing of email attachments is a critical component in preventing data leakage. Another important part of this component is the managing of the applications that can run within the container.

- Data Sharing as shown in Fig. 5.9, checks to see if functions like cut/paste are permitted across all applications or allowed across the applications within the container.
- Attachments function as shown in Fig. 5.9, checks to see if attachments are allowed to be saved, viewed within an external viewer and also checks to see if what viewers are registered. In this case, this particular function is very dependent on what file types are supported within the container from an email client perspective or a browser perspective. In many cases, external viewers are used and the important issue here is that the viewer must not allow saving the file onto another part of the device that is not part of the container.

5.7 Summary

This hybrid mobile application container solution presented here has shown itself to be very effective delivering increased application deployments, application management and improved application security; thereby, delivering an improved enterprise mobile ecosystem. Furthermore, due to its lightweight, secure and extensible design, along with having cross platforms support, it lends itself for reduced development lifecycle time and faster deployment within an enterprise.

Chapter 6
Mobile Virtualization Container Performance Analysis

In the previous chapter we presented a case study on how mobile virtualization affects applications from the user perspective where the results were directly related to the user's perception of performance, usability and security. In this chapter, through benchmark analysis, the effects of three mobile virtualization technologies are compared to the reference benchmark application written in the Mobile Virtualization Container (MVC) model which executes a several performance functions for a predefined amount of time. The mobile virtualization technologies compared to the MVC model were Divide from Enterproid, Red Bend and Blackberry Balance 10.

6.1 Performance Benchmark Analysis

A reference application with a set of benchmark tests was developed that uses the Mobile Virtualization Container architecture. These tests were designed to cover a variety of functions to test for performance impacts in data encryption and networking communication which are two of the performance metrics that are most noticeable on applications by users. In order to provide a relevant comparison, the application type selected was a hybrid application which is a mobile web application with a native application shell. This technology, as previously described, is a Cordova application running within an IBM Worklight environment where the application is developed using HTML5, CSS3 and JavaScript with the ability to use native plug-ins to access device specific functions. The main advantage on this type of application model is because the applications are fairly portable across various platforms like iOS, Android, BlackBerry and others.

IBM Worklight is an eclipse based framework that provides a light weight backend server that gives the ability to have adapters for the client application. These adapters can be used for normalizing web services in such a way that the client

D. Jaramillo et al., *Virtualization Techniques for Mobile Systems*,
Multimedia Systems and Applications, DOI 10.1007/978-3-319-05741-5_6,
© Springer International Publishing Switzerland 2014

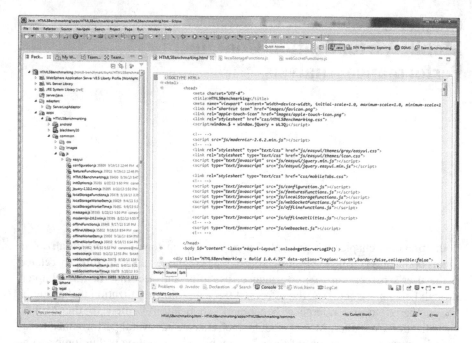

Fig. 6.1 Eclipse project performance benchmark application

doesn't need to be updated when the service changes (Fig. 6.1). In the eclipse environment the code is structured in several sections:

- Adapter code that gets deployed to the server—in this case the logging routines from the testing results get sent through an adapter and then gets logged in the backend server.
- Common code runs across all the platforms and contains the HTML, CSS and JavaScript code.
- Platform specific code contains the native specific components, also referred to as custom plug-ins. There is a directory for Android, iPhone and every other platform that the project supports.

6.2 Benchmark Components

Benchmark application was designed to be a hybrid application that implements the Mobile Virtualization Container (MVC) methodology so it simulates a contained application that can be executed outside and inside the mobile virtualization environment (Fig. 6.2).

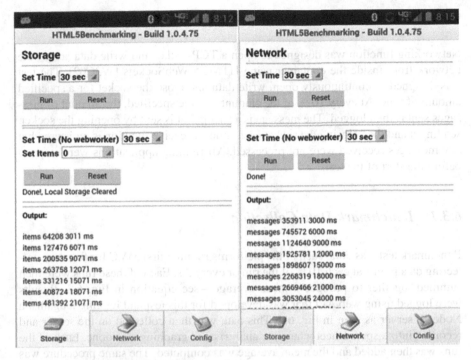

Fig. 6.2 MVCI performance benchmark application

Components tested will be using HTML5 features and are broken down into four groups:

- Storage—this will use the HTML5 localstorage feature [W3C01] to store json pairs of data inside the web container.
- Network—this will use the HTML5 websockets feature [W3C02] to communicate to an external server from within the container.
- Logging—component developed to collect the benchmark metrics with a logging framework using HTTP post requests to a simple Node JS server.

6.2.1 Storage Benchmark

Storage function was designed to store data inside the container with the HTML5 function localStorage [W3C01]. The function continuously writes data for a specified amount of time. At every 10 % of the amount of time specified, the amount of items written is then logged. The data written is a 128 bit value and it is written using the setItem method. A reset function was written so that each time the test was executed, the environment was cleared so that there was a known starting point for both memory and storage perspective. All running applications were stopped before the start of the test.

6.3   Networking Benchmark

Networking function was designed to open a TCP socket and write data across the network from inside the container using HTML5 WebSockets [W3C02]. The test was designed to continuously open, write data and close the socket for a specified amount of time. At every 10 % of the amount of time specified, the amount of messages sent is then logged. The message is a string that is sent by opening the socket, sending it and then closing the socket each time. A receive function is registered but any messages received were not processed. All running applications were stopped before the start of the test.

6.3.1   Benchmark Data Collection

Benchmark test was done each of the platforms running first MVC for five runs collecting data points at each 10 % of the run or every 3 s. Each of these runs were then summed together to produce a mean average—see equation in Fig. 6.3. Results were logged using web API that was developed for this test and logging it against a Node JS server as seen in Fig. 6.4. This data was then collected on the server and recorded into a spreadsheet where the analysis and graphing was done. Each of the runs was then added and the mean average was computed. The same procedure was

Fig. 6.3 Mean average for
benchmark results

$$R_{avg} = \frac{\sum_{i=1}^{n} r_i}{n}$$

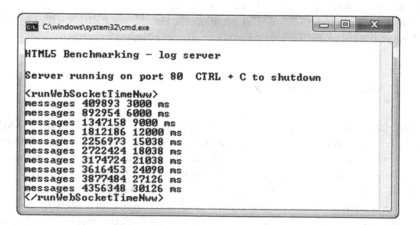

Fig. 6.4 Sample logging of execution results

then run again with the same application running the various technologies being compared, and the results were compared and graphed to produce a view that compares the difference between the technologies.

6.4 MVC Benchmark Results

The performance data was collected on each of the platforms and the results was graphed on a per test and platform basis. Basically, the MVC Storage and Networking tests were run and compared to each of the virtualization technologies Enterproid, Red Bend and Blackberry to establish if and how much overhead there is in running the applications under these mobile virtualization technologies.

6.4.1 Storage Performance Results

Storage presented some very interesting challenges because the localStorage API used behaves differently across the various platforms tested. There is a dependency on the browser level that is on the platform, which is a local storage limit of 4 MB of space on both the Red Bend and Blackberry platforms but not on iOS. As a result, additional code was incorporated to account for this limitation; and when the storage limit was reached, the storage was cleared and then continued recording more items. The time required to clear the storage was not included into the overall length of time being recorded. The metric used here was the number of items written per unit of time.

6.4.1.1 MVC vs. Enterproid on iPhone 5

Analysis of the results in Fig. 6.5 show that there is a fairly constant delta in the number of items processed between MVC and Enterproid at each time interval with a range of 4.6–6 % with an average overhead of 5.1 % and a standard deviation of 0.5 % when running for 30 s under the Enterproid container compared to the standard MVC on the iPhone 5 iOS 6.1 platform (Table 6.1).

6.4.1.2 MVC vs Enterproid on Samsung S4: Android 4.2.2

Analysis of the results in Fig. 6.6 show that there is a fairly constant delta in the number of items processed between MVC and Enterproid at each time interval with a range of 2.5–6.7 % with an average overhead of 4.6 % and a standard deviation of 1.6 % when running for 30 s under the Enterproid container compared to the standard MVC on the Samsung S4 Android platform.

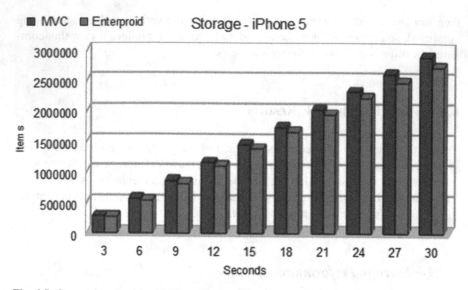

Fig. 6.5 Storage benchmark—MVC vs. Enterproid on iPhone 5

Table 6.1 Sample data collection for iPhone 5 with MVC

iPhone 5 (s)	MVC items					
	r1	r2	r3	r4	r5	Mean
3	186,286	179,630	179,403	181,031	182,925	181,855
6	372,256	358,905	359,623	362,935	365,334	363,811
9	556,405	537,556	538,388	543,719	545,737	544,361
12	740,175	715,967	718,830	723,285	726,827	725,017
15	922,830	895,670	899,585	904,770	908,513	906,274
18	1,103,971	1,074,892	1,079,608	1,085,994	1,089,206	1,086,734
21	1,286,590	1,253,172	1,259,101	1,268,037	1,270,746	1,267,529
24	1,468,848	1,430,698	1,439,470	1,449,606	1,450,721	1,447,869
27	1,650,702	1,608,748	1,615,815	1,629,922	1,631,643	1,627,366
30	1,833,434	1,786,505	1,794,314	1,810,628	1,811,084	1,807,193

6.4.1.3   MVC vs. Red Bend on Samsung S3: Android 4.0.4

Analysis of the results in Fig. 6.7 show that there is a fairly constant delta in the number of items processed between MVC and Red Bend at each time interval with a range of 12–16.8 % with an average overhead of 14.9 % and a standard deviation of 1.6 % when running for 30 s under the Red Bend hypervisor compared to the standard MVC on the Samsung S3 Android platform.

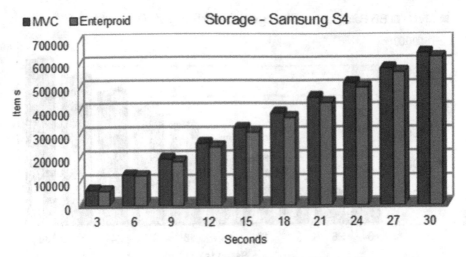

Fig. 6.6 Storage benchmark—MVC vs. Enterproid on Samsung S4

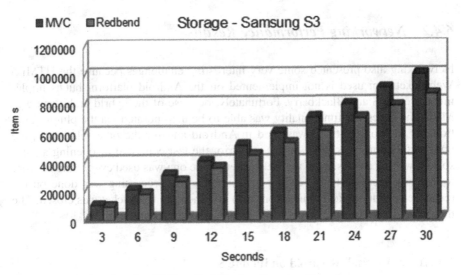

Fig. 6.7 Storage benchmark—MVC vs. Red Bend on Samsung S3

6.4.1.4   MVC vs. Blackberry Balance on Blackberry Z10

Analysis of the results in Fig. 6.8 show that there is a fairly constant delta in the number of items processed between MVC and Blackberry Balance at each time interval with a range of 0.6–3.3 % with an average overhead of 2.2 % and a standard deviation of 0.8 % when running for 30 s under Blackberry Balance compared to the standard MVC on the Blackberry Z10 platform.

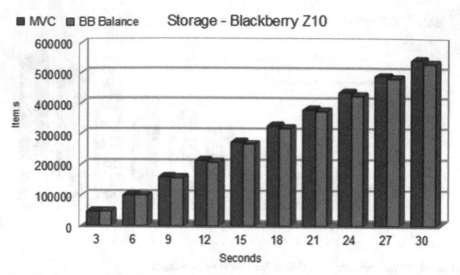

Fig. 6.8 Storage benchmark—MVC vs. Blackberry balance

6.4.2 Networking Performance Results

Networking also presented some very interesting challenges because the HTML5 Websockets api used is not implemented on the Android platform but is implemented on iOS and Blackberry. Fortunately, because of the hybrid architecture of MVC, the necessary functionality was able to be incorporated via the plugin functionality in Worklight and was used in Android and for the other platforms, the built-in functions was used. In this scenario, the test consisted of opening a websocket, writing a message to the websocket (same one was used every time in order to keep the test constant), reading the response (no processing was done on the response in order to keep things constant) and then the websocket was closed. The metric used here was the number of message written per unit of time.

6.4.2.1 MVC vs. Enterproid on iPhone 5

Analysis of the results in Fig. 6.9 show a fairly constant delta in the number of messages processed at each time interval with a average range difference of 1.67–1.88 % with an average overhead of 1.73 % and a standard deviation of 0.07 % when running for 30 s under Enterproid compared to the standard MVC on iPhone 5 platform.

6.4.2.2 MVC vs. Enterproid on Samsung S4: Android 4.2.2

Analysis of the results in Fig. 6.10 show a fairly consistent delta in the number of messages processed at each time interval with a average range difference of 2.72– 6.27 % with an average overhead of 3.43 % and a standard deviation of 1.07 %

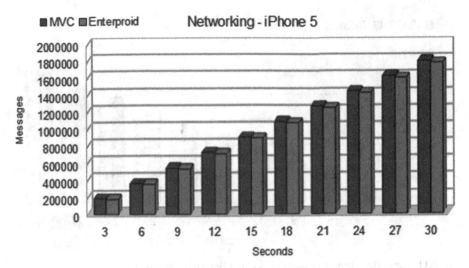

Fig. 6.9 Networking benchmark—MVC vs. Enterproid on iPhone 5

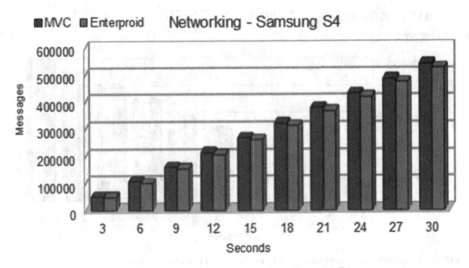

Fig. 6.10 Networking benchmark—MVC vs. Enterproid on Samsung S4

when running for 30 s under Enterproid compared to the standard MVC on the Samsung S4 Android 4.2.2 platform.

6.4.2.3 MVC vs. Red Bend on Samsung S3: Android 4.0.4

Analysis of the results in Fig. 6.11 show a constant delta in the number of messages processed at each time interval with a average range difference of 12.26–18.46 % with an average overhead of 13.95 % and a standard deviation of 1.95 % when

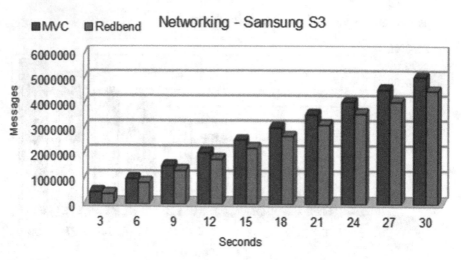

Fig. 6.11 Networking benchmark—MVC vs. Red Bend on Samsung S3

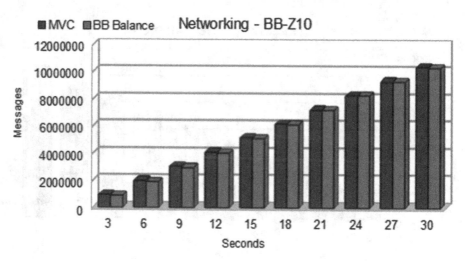

Fig. 6.12 Networking benchmark—MVC vs. Blackberry balance

running for 30 s under Red Bend compared to the standard MVC on the Samsung S3 Android 4.0.4 platform.

6.4.2.4   MVC vs. Blackberry Balance on Blackberry Z10

Analysis of the results in Fig. 6.12 have a significant delta between MVC and BB Balance for the first 15 s and then levels off to a very small delta for the second half of the 30 s period. The number of messages processed at each time interval had an average range difference of 0.02–8.55 % with an average overhead of 2.42 % and a standard deviation of 2.94 % when running for 30 s under Blackberry Balance compared to the standard MVC on the Blackberry Z10 platform.

6.5 Summary of Results

Analysis of the data provides some very interesting results that give a quantitative view of the cost of performance when running an application in a virtualized environment. In this experimentation, two attributes were recorded—storage and networking. These were chosen because these are two that are directly of importance in order to maintain a secure container so that no data is leaked or left unprotected. As a result, these virtualization technologies address this by encrypting the data in the case of storage or securing the data traffic in the case of networking.

In order to analyze the data, the results showed drastic variations when compared to each other and this was due to the various differences in the platform or functional support for the attribute being tested. As described previously, the various versions hybrid environment have different versions of HTML5 support and alternative functional compatibility had to be used or introduced. For example, on storage, Android 4.0.4 and Blackberry had limited storage compared to that in Android 4.2.2 and iOS. Another example, on networking, Android had to be supplemented with a plugin in order to provide the websocket compatibility in the hybrid container. As a result, the actual metric data could only be compared between MVC and the corresponding platform and not across all the platforms. In order to compare the results across all the platforms, the data results had to be normalized in such a way that results were independent of the platform restrictions. Since the results showed a difference in performance, that difference was then computed and tabulated as well as the average difference and its standard deviation. Once this was done, it was now possible to see what the performance impact and cost was across the different platforms each compared to that of the MVC.

Analyzing the results show that there is a correlation in performance impact across the virtualization technologies studied. From the storage perspective, Table 6.2 shows how the Blackberry platform provides least performance impact and Red Bend has the most; whereas Enterproid on both iPhone and Android are fairly comparable, at least <1 % of each other that could be attributed to hardware

Table 6.2 Mobile virtualization container platform comparison

Platform	Storage			Networking		
	Δ avg	Δ %avg	σ %avg	Δ avg	Δ %avg	σ %avg
Enterproid iPhone 5	77,438	5.13 %	0.48 %	16,680	1.73 %	0.07 %
Enterproid Samsung S4	14,324	4.62 %	1.55 %	8,948	3.43 %	1.07 %
Red Bend Samsung S3	72,759	14.89 %	1.58 %	321,228	13.95 %	1.95 %
Blackberry Balance—Z10	6,889	2.24 %	0.76 %	61,138	2.42 %	2.94 %

and OS differences. From the networking perspective, Blackberry is still within the same range as in the Storage results, but in this case Enterproid on iPhone 5 performs better than the rest. The better results on Enterproid iPhone 5 compared to that on the Android version could be attributed to the fact that the hybrid environment on iOS has built in websockets support where as on Android, an add-on plug-in had to be provided. Blackberry came in second and that could be attributed to the fact that all networking traffic in the enterprise network flows through a Mobile Data Services (MDS) gateway and hence the added latency. Red Bend had a similar rage to that as of Storage, which implies that the performance impact could be attributed to the nature of the hypervisor layer which provides a clean separation between personal and enterprise but it does it a higher cost.

In summary, the performance analysis show that there is definitely a performance impact in each of the various virtualization solutions compared to that of the Mobile Virtualization Container reference architecture and the added security and separation comes at a cost in performance.

Chapter 7
Conclusion, Contributions and Future Work

7.1 Conclusion

The explosion of mobile devices in the consumer space has been quite a disruptor for the mobile enterprise where corporate managed platforms are well defined, well contained and fairly secure. However, now with the emergence of the consumer mobile devices from Apple iOS and Google Android, the enterprise has been forced to make functional tradeoffs in order to maintain platform and data security. Due to the corporate security decisions, the end user gives up a significant amount of personal freedom and ease of use of their device. Enterprise security requirements like password complexity, device encryption, network restrictions and other techniques that restrict the access to information on the mobile device tends to drive users away and/or encourages users to find other less secure alternatives that will eventually compromise enterprise data and access. A number of mobile virtualization technologies have been presented here, each of them delivering specific features that focus on ensuring the security of the enterprise container and applications. Application level containers, type 1 and 2 hypervisors and the new hybrid application container architecture all lack the organic integration of enterprise & personal personas found in solutions like the emerging BlackBerry 10 Balance platform. Hypervisor technologies show promise; however, it is specific to the Android platform, has higher end hardware requirements and not likely to be seen on the iOS platform anytime soon.

7.2 Contributions

The work presented here focuses on the study of various techniques in mobile virtualization and how it is applied in the mobile enterprise. These contributions can be summarized as follows:

1. A comparative study of various mobile virtualization techniques and technologies has been performed.

D. Jaramillo et al., *Virtualization Techniques for Mobile Systems*,   67
Multimedia Systems and Applications, DOI 10.1007/978-3-319-05741-5_7,
© Springer International Publishing Switzerland 2014

2. A case study is presented of a mobile virtualization pilot that was conducted where one of the technologies tested and surveyed with a group of mobile users within an enterprise. The pilot survey results were then analyzed and presented.
3. An innovative reference architecture for mobile virtualization has been developed and presented, which is suitable for mobile hybrid based applications.
4. The reference architecture has been tested and compared with other mobile virtualization techniques using a performance benchmark. The results were then analyzed through the use of statistical analysis to help identify the performance impacts in each of the virtualization technologies across the various platforms that were tested.

With the work presented here, deployment plans and strategies can be developed more intelligently for integrating the growing BYOD community into the enterprise mobile ecosystem. The number of BYOD users is growing rapidly due to the rapid adoption of smartphones by consumers and with the research presented here, enterprises are able to make much more intelligent decisions in the choices of technologies to adopt and deploy.

7.3   Future Work

BYOD in the enterprise is not going away anytime soon and there is huge interest for the enterprises to increase the adoption of this community. Due to enterprise policies, the BYOD adoption of the enterprise is not going as quick as expected or desired. As presented in the background work, there are many technologies of various types that try to fill the needs for the BYOD community, but very few of them work cleanly across the different platforms and the architecture presented here allows the enterprise to containerize its applications and interface the necessary mobile device management systems such that both world can be possible. There are several aspects about the Mobile Virtualization Container presented here that can be extended into future work. One such function would be to extend the MVC Interfaces into a SDK that would be used into native mobile applications. Another piece of work would be to extend this container as a mobile web browser such that server based mobile web applications could also be containerized. Both of these are not small pieces of work and are projects that various companies in the market place are tackling. On the same token, the device/OS manufactures are also rapidly incorporating many of the features mentioned here into the base OS or building the underpinnings into the hardware itself to support these technologies.

References

[AlLu01] V. G. Sarah Allen and L. Lundrigan. "Pro Smartphone Cross-Platform Development." Apress, 2010.

[AnDa01] Jeremy Andrus, Christoffer Dall, Alexander Van't Hof, Oren Laadan, and Jason Nieh. "Cells: A Virtual Mobile Smartphone Architecture." Proceedings of the 23rd ACM Symposium on Operating Systems Principles (SOSP 2011). Portugal, Cascais. 173–87. Web.

[BaBu01] Ken Barr, Prashanth Bungale, Stephen Deasy, Viktor Gyuris, Perry Hung, Craig Newell, Harvey Tuch, and Bruno Zoppis. "The VMware mobile virtualization platform: is that a hypervisor in your pocket?" *SIGOPS Oper. Syst. Rev.* 44, 4 (2010), 124–135.

[BaDr01] Paul Barham, Boris Dragovic, Keir Fraser, Steven Hand, Tim Harris, Alex Ho, Rolf Neugebauer, Ian Pratt, Andrew Warfield. "Xen and the art of virtualization." Proceedings of the nineteenth ACM symposium on Operating systems principles, October 19–22, 2003, Landing, NY, USA

[BaTe01] "Balance Technology." BlackBerry - BlackBerry Balance Technology Separates Personal from Business Information. RIM, n.d. Web. 14 Feb. 2012. <http://us.blackberry.com/business/software/blackberry-balance.html>.

[Bran01] John Brandon. "Is Mobile Virtualization Ready for Your Business?" CIO. N.p.,15 Mar. 2012. Web. 11 Apr. 2012. <http://www.cio.com/article /702299/Is_Mobile_Virtualization_Ready_for_Your_Business_>.

[BrDr01] Jörg Brakensiek, Axel Dröge, Martin Botteck, Hermann Härtig, and Adam Lackorzynski. "Virtualization as an enabler for security in mobile devices" *Proceedings of the 1st workshop on Isolation and integration in embedded systems* (IIES '08), Michael Engel and Olaf Spinczyk (Eds.). ACM, New York, NY, USA, 17–22, 2008

[BuDa01] Sven Bugiel, Lucas Davi, Alexandra Dmitrienko, Stephan Heuser, Ahmad-Reza Sadeghi, Bhargava Shastry. "Practical and lightweight domain isolation on Android." Proceedings of the 1st ACM workshop on Security and privacy in smartphones and mobile devices, October 17–17, 2011, Chicago, IL, USA

[9] Jaymar Cabebe. "Android 4.2 Adds Multiple Users and Panoramic Photos, Copies Swype and AirPlay." CNET. N.p., 29 Oct. 2012. Web. 02 Nov. 2012. <http://reviews.cnet.com/8301-19736_7-57542145-251/android-4.2-adds-multiple-users-and-panoramic-photos-copies-swype-and-airplay/>.

[10] "Cells: Lightweight Virtual Smartphones." *Cells: Lightweight Virtual Smartphones*. Columbia University Department of Computer Science, n.d. Web. 23 May 2012. <http://systems.cs.columbia.edu/projects/cells>.

[Cell02] "The ThinVisor Mobile Device Virtualization Architecture." Cellrox, Nov 2011. http://www.cellrox.com/wp-content/uploads/2012/02/Cellrox-ThinVisor-Architecture.pdf

D. Jaramillo et al., *Virtualization Techniques for Mobile Systems*,
Multimedia Systems and Applications, DOI 10.1007/978-3-319-05741-5,
© Springer International Publishing Switzerland 2014

[CoCh01] Landon P. Cox, Peter M. Chen. "Pocket Hypervisors: Opportunities and Challenges." Proceedings of the Eighth IEEE Workshop on Mobile Computing Systems and Applications, p.46–50, March 08–09, 2007

[Cox1] John Cox. "MobileIron Extends Control over Enterprise IOS Apps." Network World. N.p., 9 Dec. 2010. Web. 02 Nov. 2012. <http://www.networkworld.com/news/2010/120810-mobileiron-4.html>.

[Cree01] Mache Creeger. ACM CTO Roundtable on Mobile Devices in the Enterprise. August 3, 2011. http://queue.acm.org/detail.cfm?id=2016038

[CrSo01] Stacy K. Crook, Ian Song. "Mobile Virtualization Technology Assessment." IDC, May 2011. Web. 02 Oct. 2012. <http://www.idc.com/getdoc.jsp?containerId=228088>.

[DaNi01] Christoffer Dall, Jason Nieh. "KVM for ARM." Proceedings of the 12th Annual Linux Symposium, Ottawa, Canada, July 13–16, 2010.

[Dev01]"Device Administration." Google Android SDK, October 20, 2012. http://developer.android.com/guide/topics/admin/device-admin.html

[Dolc01] Jessica Dolcourt. "Divide for Android Takes on BlackBerry, Sprint ID." CNET. N.p., 28 Feb. 2011. Web. 22 June 2011. <http://www.cnet.com/8301-17918_1-20036669-85.html>.

[Ente01] "Implementing Your BYOD Mobility Strategy." The Divide Platform Enables BYOD Mobility. Enterproid, 2012. Web. 14 Oct. 2012. <http://www.divide.com/>.

[Epst01] Zach Epstein. "RIM Announces BlackBerry Balance for Work-life Balance on a Single Smartphone." BGR: The Three Biggest Letters In Tech. BGR, 2 May 2011. Web. 14 Feb. 2012. <http://www.bgr.com/2011/05/02/rim-announces-blackberry-balance-for-work-life-balance-on-a-single-smartphone/>.

[FaNa01] Keith I. Farkas, Chandra Narayanaswami, Jason Nieh. "Guest Editors' Introduction: Virtual Machines." Pervasive Computing, IEEE, vol.8, no.4, pp.6–7, Oct.-Dec. 2009.

[Good01] "Balancing Security and Speed: Developing Mobile Apps for Enterprise." Good Dynamics White Paper. Good Technology, 2011. Web. 27 Nov. 2011. <http://media.www1.good.com/documents/good_dynamics_wp.pdf>.

[Gohr01] Nancy Gohring. "Red Bend Working on Mobile Virtualization." CIO. N.p., 12 Oct. 2011. Web. 11 Apr. 2012. <http://www.cio.com/article/691671/ Red_Bend_Working_on_Mobile_Virtualization>.

[Gart01] Phillip Redman, John Girard, Terrence Cosgrove, Monica Basso "Magic Quadrant for Mobile Device Management Software." 23 May 2013 ID:G00249820 http://www.gartner.com/technology/reprints.do?id=1-1FRG59X&ct=130523&st=sbJ

[Grum01] Galen Gruman. Mobile and BYOD Deep Dive. Rep. N.p.: InfoWorld, November 2011. http://www.infoworld.com/d/mobile-technology/download-the-byod-and-mobile-strategy-deep-dive-179850/>.

[Grum02]Galen Gruman. Mobile Device Management Deep Dive. N.p.: InfoWorld, March 2011. <<http://www.infoworld.com/d/mobilize/mobile-management-infoworlds-expert-guide-371-0/>.

[GuPi01] Kevin Gudeth, Matthew Pirretti, Katrin Hoeper, Ron Buskey. "Delivering secure applications on commercial mobile devices: the case for bare metal hypervisors." Proceedings of the 1st ACM workshop on Security and privacy in smartphones and mobile devices, October 17–17, 2011, Chicago, Illinois, USA

[Hale01] Ronen Halevy. "RIM Finally Details Features of BlackBerry Balance." BerryReview, 3 May 2011. Web. 14 Feb. 2012. <http://www.berryreview.com/2011/05/03/rim-finally-details-features-of-blackberry-balance/>.

[Haza01] John Hazard. "Enterproid Divides Work and Personal on Android Devices, Fires at BlackBerry." ZDNet. Between the Lines, 28 Feb. 2011. Web. 22 June 2011. <http://www.zdnet.com/blog/btl/enterproid-divides-work-and-personal-on-android-devices-fires-at-blackberry/45407>.

[Heis01] Gernot Heiser. The Motorola Evoke AQ4 – A Case Study in Mobile Virtualization. Open Kernel Labs, 2009 – okl4.net. http://www.ok-labs.com/_assets/image_library/evoke.pdf

[Heis02] Gernot Heiser. 2011. "Virtualizing embedded systems: why bother?" *Proceedings of the 48th Design Automation Conference* (DAC '11). ACM, New York, NY, USA, 901–905.

[IBM01] "Are you ready for BYOD? Here are seven questions you should answer as you roll out new mobile capabilies." IBM Software – Thought Leadership White Paper, December 2011.

[JaCo01] David Jaramillo, Thomas Cook, Neil Katz, William Bodin, Simon Cooper, Craig Becker, Robert Smart, Charisse Lu "Mobile Innovation Applications for the BYOD Enterprise User" IBM Journal of Research and Development, Vol. 57, No 6, Paper 6, November/December 2013.

[JaFu01] David Jaramillo, Borko Furht, Ankur Agarwal "Virtualization Techniques for Mobile Devices" International Review on Computers and Software (IRECOS) Vol 8, No 8, August 2013.

[JaKa01] David Jaramillo, Neil Katz, Bill Bodin, William Tworek, Robert Smart, Thomas Cook "Cooperative Solutions for Bring Your Own Device (BYOD)" IBM Journal of Research and Development, Vol. 57, No. 6, Paper 5, November/December 2013.

[JaNe01] David Jaramillo, Richard Newhook, Robert Smart. "Cross-platform, secure message delivery for mobile devices." Southeastcon, 2013 Proceedings of IEEE, vol., no., pp.1,5, 4–7 April 2013.

[JaUg01] [David Jaramillo, Viney Ugave, Charisse Lu, Rick Alther. "Android OS in the Enterprise." Software Developer's Journal, Vol 2, No 11, pp.64–71, 2013.

[JaSm01] David Jaramillo, Robert Smart, Borko Furht, Ankur Agarwal. "A secure extensible container for hybrid mobile applications." Southeastcon, 2013 Proceedings of IEEE, vol., no., pp.1,5, 4–7 April 2013.

[KhNa01] S. Khan, M. Nauman, A. T. Othman, S. Musa. "How secure is your smartphone: An analysis of smartphone security mechanisms." Cyber Security, Cyber Warfare and Digital Forensic (CyberSec), 2012 International Conference on, vol., no., pp.76–81, 26–28 June 2012.

[Khar01] Olga Kharif. "Virtualization Goes Mobile." Businessweek - Business News, Stock Market & Financial Advice. N.p., 22 Apr. 2008. Web. 22 Sept. 2012. <http://www.businessweek.com/stories/2008-04-22/virtualization-goes-mobilebusinessweek-business-news-stock-market-and-financial-advice>.

[LaWa01] Adam Lackorzynski, Alexander Warg, "Taming subsystems: capabilities as universal resource access control in L4." Proceedings of the Second Workshop on Isolation and Integration in Embedded Systems, p.25–30, March 31–31, 2009, Nuremburg, Germany.

[LaLi01] Matthias Lange, Steffen Liebergeld, Adam Lackorzynski, Alexander Warg, Michael Peter. "L4Android: a generic operating system framework for secure smartphones." Proceedings of the 1st ACM workshop on Security and privacy in smartphones and mobile devices, October 17–17, 2011, Chicago, Illinois, USA

[LeSu01] Sung-Min Lee, Sang-Bum Suh, Jong-Deok Choi. "Fine-grained I/O access control based on Xen virtualization for 3G/4G mobile devices." Proceedings of the 47th Design Automation Conference, June 13–18, 2010, Anaheim, CA.

[Madd01] Jack Madden. "BYOD Smackdown 2012: Enterproid Divide Creates a Secure Work Persona on Personal Devices." *ConsumerizeIT*. N.p., 16 Feb. 2012. Web. 10 Oct. 2012. <http://www.consumerizeit.com/blogs/consumerization/archive/2012/02/16/byod-smackdown-2012-enterproid-divide-creates-a-secure-work-persona-on-personal-devices.aspx>.

[Madd02] Jack Madden. "Cellrox Offers Multiple "personas" for Android Phones. Is This How We Wish VMware Horizon Mobile Worked?" Brianmadden.com. N.p., 22 Sept. 2012. Web. 29 Sept. 2012. <http://www.brianmadden.com/blogs/jackmadden/archive/2012/09/21/cellrox-offers-multiple-personas-for-android-phones-is-this-how-we-wish-vmware-horizon-mobile-worked.aspx>.

[MaDo01] Stuart E. Madnick, John J. Donovan, Application and analysis of the virtual machine approach to information system security and isolation, Proceedings of the workshop on virtual computer systems, p.210–224, March 26–27, 1973, Cambridge, Massachusetts, United States

[Mess01] Ellen Messmer. "Security Minefield: BYOD Will Bedevil IT Security in 2012." InfoWorld. Network World, 21 Dec. 2011. Web. 22 Sept. 2011. <http://www.infoworld.com/d/security/security-minefield-byod-will-bedevil-it-security-in-2012-182348?source=IFWNLE_nlt_wrapup_2011-12-21>.

[Mobi01] "Building "Bring-Your-Own-Device" (BYOD) Strategies." BYOD Strategies White Paper. MobileIron, 2011. Web. 27 Nov. 2011. <http://info.mobileiron.com/BYOD_1.html>.

[Morg01] Timothy P. Morgan. "Xen Hypervisor Ported to ARM Chips." Xen Hypervisor Ported to ARM Chips. The Register, 11 Nov. 2011. Web. 11 Dec. 2011. <http://www.theregister.co.uk/2011/11/30/xen_kvm_hypervisor_for_arm_chips/>.

[OmSa01] A. Omezzine, Sami Yangui, N. Bellamine, S. Tata. "Mobile Service Micro-containers for Cloud Environments." Enabling Technologies: Infrastructure for Collaborative Enterprises (WETICE), 2012 IEEE 21st International Workshop on, vol., no., pp.154–160, 25–27 June 2012.

[Open01] Open Kernel Labs. OKL4 Microvisor, Mar. 2011. <http://www.ok-labs.com/products/okl4-microvisor>.

[Perl01] Jason Perlow. "Android Virtualization: It's Time." ZDNet, 19 Jan. 2011. Web. 23 Aug. 2011. <http://www.zdnet.com/blog/perlow/android-virtualization-its-time/15588>.

[Phon01] PhoneGap. http://phonegap.com, 2012.

[Rama01] Rahul Ramasubramanian. "Exploring Virtualization Platforms for ARM based Mobile Android Devices." Master Thesis, North Carolina State University, 2011. http://www.lib.ncsu.edu/resolver/1840.16/6998

[Red01] Red Bend Software. Mobile Virtualization, 2011. http://www.redbend.com

[ReGi01] Phillip Redman, John Girard, Monica Basso. "Magic Quadrant for Mobile Device Management Software." N.p., 17 May 2012. Web. 08 Nov. 2012. <http://www.gartner.com/id=2019515>.

[Rudo01] L. Rudolph. "A Virtualization Infrastructure that Supports Pervasive Computing." Pervasive Computing, IEEE, 8(4), 8–13, 2009.

[SaVa01] Reiner Sailer, Trent Jaeger, Enriquillo Valdez, Ramon Caceres, Ronald Perez, Stefan Berger, John Linwood Griffin, Leendert van Doorn. "Building a MAC-Based Security Architecture for the Xen Open-Source Hypervisor." Proceedings of the 21st Annual Computer Security Applications Conference, p.276–285, December 05–09, 2005.

[Secu01] "Security Minefield: BYOD Will Bedevil IT Security in 2012." n.d.: n. pag. InfoWorld. 21 Dec. 2011. Web. 22 Dec. 2011. <http://www.infoworld.com/d/security/security-minefield-byod-will-bedevil-it-security-in-2012-182348>.

[SeNa01] Balasubramanian Seshasayee, Nitya Narasimhan, Ashish Bijlani, Ankur Pai, and Karsten Schwan. 2008. "VStore: efficiently storing virtualized state across mobile devices." In Proceedings of the First Workshop on Virtualization in Mobile Computing (MobiVirt '08). ACM, New York, NY, USA, 43–47.

[StKa01] Udo Steinberg, Bernhard Kauer. "NOVA: a microhypervisor-based secure virtualization architecture." Proceedings of the 5th European conference on Computer systems, April 13–16, 2010, Paris, France

[StDR01] William Stofega, Stephen D. Drake. "Mobile Virtualization: Accelerating Innovation in Next-Generation Services." IDC, September 2012.

[Thom01] Keir Thomas. "Virtualization Boosts LG Android Phones - PCWorld Business Center. Reviews and News on Tech Products, Software and Downloads" PCWorld. N.p., 7 Dec. 2010. Web. 22 Sept. 2012. <http://www.pcworld.com/businesscenter/article/212764/virtualization_boosts_lg_android_phones.html>.

[VMTN1] "VMware End-User Computing Blog." VMware End-User Computing Blog. N.p., 30 Aug. 2011. Web. 14 Feb. 2012. <http://blogs.vmware.com/euc/2011/08/vmworld-2011-announcing-vmware-horizon-mobile-manager.html>.

[W3C01] WebStorage. http://www.w3.org/TR/webstorage/

[W3C02] WebSocket, http://www.w3.org/TR/websockets/

[Whit01] Sally Whittle. "VMware Foresees Mobile Virtualization in 2010 l Business Tech - CNET News." Technology News - CNET News. N.p., 21 May 2009. Web. 22 Sept. 2012. <http://news.cnet.com/8301-1001_3-10246338-92.html>.

[Wint01] [Wint01] Johannes Winter. "Trusted computing building blocks for embedded linux-based ARM trustzone platforms." Proceedings of the 3rd ACM workshop on Scalable trusted computing, October 31–31, 2008, Alexandria, Virginia, USA

[Xen01] Xen ARM Project, http://wiki.xen.org/wiki/Xen_ARM_(PV)

[XuZh01] Xudong Ni, Zhimin Yang, Xiaole Bai, A.C. Champion, Dong Xuan. "DiffUser: Differentiated user access control on smartphones," Mobile Adhoc and Sensor Systems, 2009. MASS '09. IEEE 6th International Conference on, vol., no., pp.1012–1017, 12–15 Oct. 2009

[ZaAc01] [ZaAc01] Xinwen Zhang, Onur Acıiçmez, Jean-Pierre Seifert. "A trusted mobile phone reference architecture via secure kernel." Proceedings of the 2007 ACM workshop on Scalable trusted computing, November 02–02, 2007, Alexandria, Virginia, USA

[Zieg01] Chris Ziegler. "Work, Play on a Single Phone: LG Teams up with VMware to Deploy Android Handsets with Virtualization." *Engadget*. N.p., 7 Dec. 2010. Web. 22, June 2011. http://www.engadget.com/2010/12/07/work-play-on-a-single-phone-lg-teams-up-with-vmware-to-deploy/

Printed in the United States
by Bookmasters

Printed in the United States
By Bookmasters